Atmospheric Pollutants in Forest Areas

Atmospheric Pollutants in Forest Areas

Their Deposition and Interception

edited by

H.-W. GEORGII

Institute for Meteorology and Geophysics, University of Frankfurt, Germany

D. REIDEL PUBLISHING COMPANY

A MEMBER OF THE KLUWER ACADEMIC PUBLISHERS GROUP

DORDRECHT / BOSTON / LANCASTER / TOKYO

Library of Congress Cataloging in Publication Data

Atmospheric pollutants in forest areas.

    Includes index.
    1. Forest ecology. 2. Air—Pollution—Environmental aspects.
I. Georgii, H. W.
QH541.5.F6A86   1986   574.5'2642   86-17736
ISBN 90-277-2317-6

Published by D. Reidel Publishing Company,
P.O. Box 17, 3300 AA Dordrecht, Holland.

Sold and distributed in the U.S.A. and Canada
by Kluwer Academic Publishers,
101 Philip Drive, Assinippi Park, Norwell, MA 02061, U.S.A.

In all other countries, sold and distributed
by Kluwer Academic Publishers Group,
P.O. Box 322, 3300 AH Dordrecht, Holland.

All Rights Reserved
© 1986 by D. Reidel Publishing Company, Dordrecht, Holland
No part of the material protected by this copyright notice may be reproduced or
utilized in any form or by any means, electronic or mechanical
including photocopying, recording or by any information storage and
retrieval system, without written permission from the copyright owner

Printed in The Netherlands

TABLE OF CONTENTS

PREFACE

DEPOSITION IN FOREST-ECOSYSTEMS

A.W.M. Vermetten, P. Hofschreuder, H. Harssema
Deposition of gaseous pollutants in a douglas fir forest — 3

G. Enders, U. Teichmann
Gasdep – Gaseous deposition measurements of SO_2, NO_x and O_3 to a spruce stand: Conception, instrumentation and first results of an experimental project — 13

K. Hanewald
Comparison of SO_2-concentrations and immission-rates with bulk precipitation data — 25

S. Grosch
Wet and dry deposition of atmospheric trace elements in forest areas — 35

H. M. Brechtel, Á. Balázs, F. Lehnardt
Precipitation input of inorganic chemicals in the open field and in forest stands – Results of investigations in the State of Hesse — 47

P. Valenta, V. D. Nguyen
Trends of heavy metal pollution by wet deposition in the Federal Republic of Germany during 1980 – 1984 — 69

G. Gietl, A. M. Rall
Bulk deposition into the catchment "Grosse Ohe": Results of neighbouring sites in the open and under spruce at different altitudes — 79

W. Kuttler
Input of atmospheric pollutants in a remote highland area — 89

G. Glatzel, M. Kazda, G. Markart
Winter deposition rates of atmospheric trace constituents in forests: Assessment of total input — 101

INTERCEPTION OF FOGWATER

F. Dröscher
Design of a fog water collector for chemical analysis — 111

G. Schmitt
The temporal distribution of trace element concentrations
in fog water during individual fog events   129

P. Winkler
Observation on fog water composition in Hamburg   143

CASE-STUDIES

R. Zimmermann
Temporal variations of trace substances during individual
rain events   155

H. Borchert
Preliminary results and experiences with a new in-situ
measurement system of rainfall acidity in forest areas
of Rheinland-Pfalz / Federal Republic of Germany   165

G. Baumbach
Occurrence of gaseous pollutants in forest stands   177

G. Fuchs, K. Bächmann
Possible influences of chlorine containing species on forests   189

EFFECTS ON ECOSYSTEMS

E. Rohbock
Watersolubility of heavy metals in deposition samples
- Interpretation and prediction of bioavailability   201

M. Kazda, G. Glatzel
Dry deposition, retention and wash-off processes of heavy metals
in beech crowns: analysis of sequentially sampled stemflow   215

A. Fangmeier, L. Steubing
Cadmium and lead in the food web of a forest ecosystem   223

H.-J. Ballach, W. Elling, H. Greven, R. Wittig
Studies on biocenoses, individual organisms and deposition rates
in the Egge mountains, an area heavily affected by forest decline   235

E. Matzner
Deposition/canopy-interactions in two forest ecosystems
of Northwest Germany   247

J. Godt, M. Schmidt, R. Mayer
Processes in the canopy of trees: internal and external
turnover of elements   263

LIST OF PARTICIPANTS   275

SUBJECT INDEX   285

PREFACE

In November 1981 a first symposium with the topics of "Acid Deposition of Atmospheric Pollutants" was organised in Oberursel/Taunus to introduce the problems and first results of research-activities on wet and dry deposition of pollutants and on acid precipitation.

In the meantime the hazard to forest and vegetation became more dramatic and research-projects to investigate the input of pollutants to forest-ecosystems have been initiated by several interdisciplinary groups. The rapidly increasing interest in the problems of forest-decay and the many open questions with respect to the diagnosis of the forest-damage were the background for the organisation of a second symposium which was held in November 1985 at the same location in Oberursel/Taunus. It was mainly concerned with new techniques of sampling and analyzing pollutants in forest areas.
Besides deposition, one important pathway of pollutants in orographic terrain is the interception of fog-droplets by vegetation. Special emphasis was laid on the chemical composition of fog. The symposium successfully assembled scientists from the field of atmospheric research with those studying the effect of pollutants on trees and vegetation in order to reduce the many open questions in connection with forest desease. The proceedings presented in this volume are a substantial contribution to the understanding of deposition and interception of pollutants in forest-areas. Thanks to the authors the volume contains a lot of new research results and presents therefore a true picture of our present knowledge.

The success of the symposium was to a great deal due to the work of my associates and I would like to thank Mr. S. Grosch, Mr. G. Schmitt and Mrs. H. Wallenwein for their assistance.

Hans-Walter Georgii

Deposition in Forest-Ecosystems

DEPOSITION OF GASEOUS POLLUTANTS IN A DOUGLAS FIR FOREST

A.W.M. Vermetten, P. Hofschreuder, H. Harssema
Agricultural University Wageningen
Department of Air Pollution
P.O. Box 8129
6700 EV Wageningen
The Netherlands

ABSTRACT. This year a Dutch acid deposition monitoring program will be started in two stands of Douglas fir. The program includes continuous measurement of concentration profiles of the main gaseous pollutants SO_2, NO_x, NH_3 and O_3 as well as monitoring of physical and chemical parameters in the soil. To evaluate the response of the trees to air pollution stress there will also be photosynthesis measurements on single branches under controlled conditions. The first results are to be expected in the last months of 1986. In this contribution the experimental lay-out is described and some possible ways of interpretation of the measurements are discussed.

1. INTRODUCTION

On the first of July 1985 a Dutch National Research Program on acid rain was established. |1| The program will run for two and a half years and will have to provide scientific information, currently unavailable, on the "Acid Rain" problem to enable policy makers to take counter measures.

Main topics of the program are: research in exposure-effect relationships, research into NH_3-emission control techniques and study of the effectiveness of control measures, this in addition to the already existing projects on SO_2 and NO_x emission control. The exposure-effect studies will focus on forests, natural vegetation and agricultural crops and comprise experimental field studies, comparative field studies and studies under controlled conditions.

The contribution of our department, a research project called 'Air Pollution in Forest Canopies', is a part of the experimental field studies together with research on tree physiology, soil physics and chemistry, root growth and research on mycorrhiza. These studies will take place simultaneously at the same plots to enhance coherence of the program and to provide other research groups with additional information.

2. AIMS OF THE PROJECT

One of the final aims of the research program on forests will be an estimate of the influence of air pollution, among other stress factors, on the vitality and growth rate of the Douglas fir (Douglas menziesii). Our part in the program will be the determination of air pollution levels above, within and below the forest canopy of two Douglas plantations, one with good and one with poor to moderate growth rate. Concentration profiles of the main gaseous pollutants SO_2, NO_x O_3 and NH_3 will be measured continuously.
As a further goal we will try to develop a model to quantify wet and dry deposition onto the trees and the soil under various meteorological conditions.

Finally we will have to determine exposure-effect relationships for the Douglas fir in regard to possible 'sensitive periods' for physiological damage and episodicity of the deposition. This will be done in close cooperation with research groups on tree physiology.

In addition to the continuous monitoring program over a period of about two years, other measurements will be taken during a few months each year. TNO (Netherlands Organisation for Applied Scientific Research) will determine levels of reactive hydrocarbons; ECN (Netherlands Energy Research Foundation) will sample acid gases and determine acids and photochemically active species in fog and dew. These measurements are intended to provide information about ohter components which can possibly play an important role in the decline of the Dutch forests.

The selection of the field stations has been rather difficult until now, due to the various requirements of the cooperating research groups, such as aerodynamic fetch, soil homogenity and stand vitality. Most Dutch forest plantations are rather small and generally surrounded by quite different environments, including industrial and agricultural areas with possible emissions of the pollutants of interest. Consequently parts of the research effort have to evaluate the influence of nearby sources, to quantify edge effects and to judge the representativity of the measuring sites.

3. EXPERIMENTAL LAY-OUT

In the central part of both Douglas stands a 30 m measuring tower will be erected with five arms bearing sensors. There will be two measuring levels above the canopy, two within and one below (see figure 1.). At all heights concentrations of SO_2, NO_x, O_3, NH_3, wind speed and direction, temperature and relative humidity will be measured. At the top level rainfall intensity and global radiation will be registered. If possible, the leaf wetness period will be monitored at the same levels.

Signal cables and Teflon tubing transport electrical signals and gases into a container next to the tower. The air entering the sampling lines is filtered from aerosols to avoid contamination of the tubing by aerosols. Filters will be renewed on a regular basis to avoid gas-

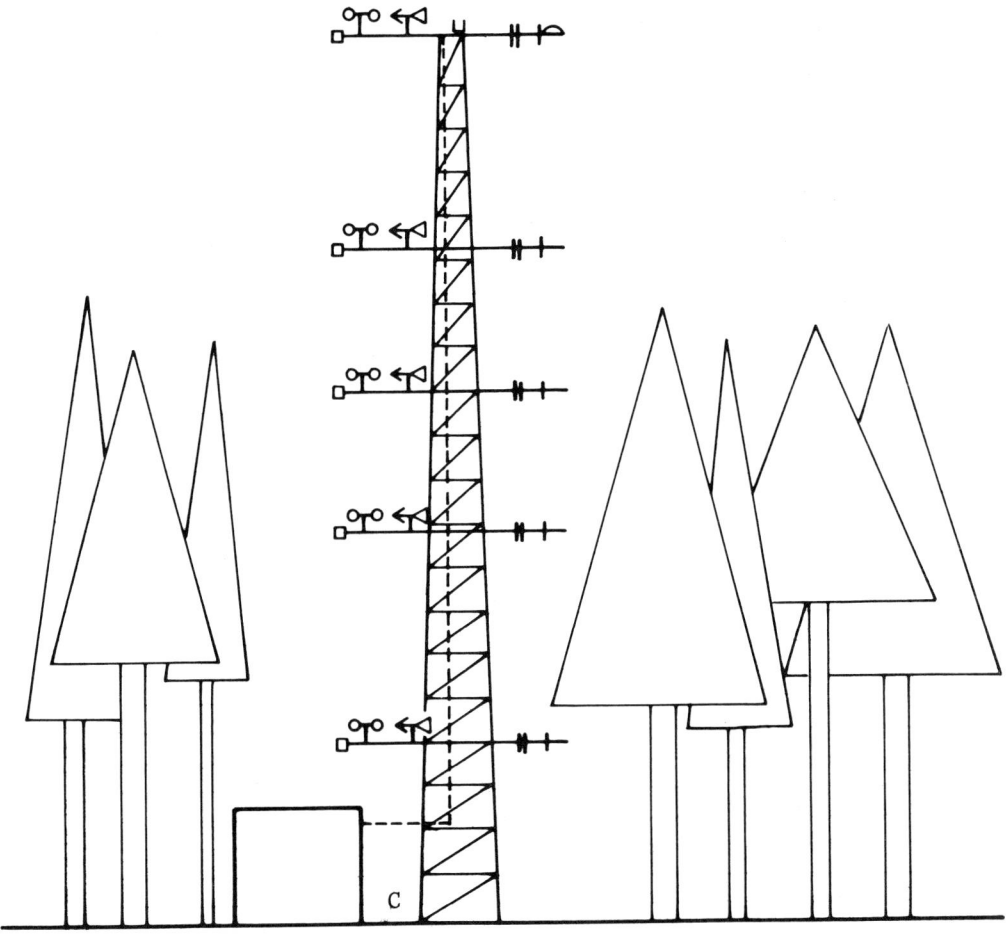

FIGURE 1. Lay-out of the measuring tower

C = Measuring container with NO_x, SO_2, NH_3 and O_3 monitors and data processing unit

◐ = incoming solar radiation intensity

| = temperature

|| = relative humidity

⊔ = rainfall intensity

◀ = wind direction

⚲ = wind speed

▫ = entrance filter for teflon tubing

particle interferences. Both tubing and filter housings will be heated slightly to keep all surfaces dry.

Within the measuring container, which is air conditioned, the instruments mentioned in table I analyse the incoming air by taking a small sidestream from the fast flowing mainstream of air. This is done to achieve a short residence time in the tubing. All other dimensions will be chosen so that an excessive pressure drop in the sampling system is avoided. Of course there will still be some sampling artifacts due to dissociation of ammonium salts on the filter, but we can avoid other severe artifacts such as differences between analysers or contamination of the tubing by retention of aerosols. The performance of the sampling system will be tested intensively in the laboratory and during the field experiments.

TABLE I. Gas analysis Equipment

Component	Analyser	Principle + Remarks
SO_2	Thermo Electron 43W	Fluorescence
$NO + NO_2$	Monitor Labs 8840 or Philips PW 9764/00	Chemoluminescence
O_3	Bendix 8002	Chemoluminescence + ethylene
NH_3	Monitor Labs 8840	Chemoluminescence + Tungsten oxide preconcentration units and SS catalyzer

Signal processing and steering of the equipment is done by an 'ARCOM'-system connected to a Digital PDP11 DTC 7367/WS computer, which also services equipment installed by other groups. The data will be stored on magnetic tape to obtain a central database, which can be used by all participants.

All instruments will be guarded against lightning as much as possible. For example, measuring towers and containers will be grounded thoroughly; for the incoming signals optical couplers will be used.

In addition to the continuous measurements our laboratory will organize measuring campaigns to characterize aerosol size-distribution and composition near the field site.

With the help of a mobile tower we will try to quantify the spatial variations in the concentrations of gaseous pollutants, especially near forest edges.

We are still considering direct measurement of deposition fluxes by exposure of surrogate surfaces. Due to differences in micrometeorological, physical and chemical properties between these and natural surfaces it will only give us an idea of the spatial variations in aerosol deposition and not the absolute amounts.

As we have already mentioned, ECN and TNO will sample acid gases, dew, fog and reactive hydrocarbons.

Bulk precipitation, wet-only precipitation, throughfall and litterfall will be sampled and analysed by the Department of Soil Science and Geology of the Agricultural University of Wageningen, in cooperation with the Laboratory of Physical Geography and Soil Science of the University of Amsterdam. They will also determine the soil water content, soil saturation pressures and soil temperature, as well as the chemical composition of the soil solution and the soil air.

Tree growth and leaf area index will be measured regularly by the Insitute for Forestry and Landscape Planning 'De Dorschkamp' (Wageningen). Needles of different ages and exposure to radiation will be analysed for composition. The Dorschkamp will also perform photosynthesis measurements in transparant cuvettes which will be mounted around twigs of a few selected trees. These measurements have to provide the link between the average gas concentrations and the direct effects on the needles. Therefore the twigs will be exposed to filtered air, (polluted) ambient air and air with a fixed level of one or more main pollutants (SO_2, NO_x, NH_3 or O_3). CO_2 and water vapour concentrations will be measured at the exit and entrance of the cuvettes.

Root growth will be monitored by another department of the Agricultural University (Forestry), using an endoscopic technique in prepunched holes in the forest floor. After the completion of the monitoring program some destructive samples of roots and trees will be taken.

4. INTERPRETATION

During the monitoring program a very large dataset will gradually become available. Analysis will not be easy, since many external factors can influence growth and vitality of a tree. A first step in determining the role of air pollution in this process will be the statistical evaluation of the concentration levels, e.g. calculating average concentration levels, frequency distributions, and their relation to the large scale meteorology above the forest.

As a next step we will try to estimate the pollutant fluxes to parts of the vegetation and the soil from the measured concentration profiles. A rough method of doing so is putting these fluxes equal to the measured concentration gradients times a turbulent diffusion constant: $F_c = K_c * dC/dz$. By using the analogy of electricity (Ohm's law) we can put the pollutant flux equal to the current, the concentration gradient to a potential difference and the reciprocal value of the diffusion constant to a resistance. Then we can treat the forest as an electrical circuit and calculate the fluxes, provided that we have the correct values for the resistances.

Such a 'resistance model' can have as many layers as desirable. An example of a simple one-layer model was given in Grace et al. |2| (figure 2.). As radiation intensity, stomatal resistances, wetness of the canopy and the amount of needles and twigs per unit volume show large variations within the canopy, we intend to use a multilayer model. |3| However, this still is a rather simple model, which in its present form can only be used for the calculation of vertical fluxes in a stationary situation. To include dynamic processes as interception, evaporation, dew formation, photosynthesis rate and stomatal responses to various factors, we need a micrometeorological model such as MICROWEATHER. (Goudriaan, 1977 & 1979) |4| |5| Pollutant exchanges can be built in the same way as the CO_2 and water vapor exchange; uptake resistances for these processes can be calculated by this model. Nevertheless good care should be taken when modelling the uptake of reactive gases as HNO_3, for which stomatal resistance is essentially zero. Further exchange processes are quite different within a wet canopy, which can be a perfect sink for most gases until saturation of the water phase occurs. Therefore some of the modelling effort will concentrate on the uptake by and the chemical processes in droplets or layers of water. From laboratory experiments in a small wind tunnel and some model calculations we already know that SO_2 and NH_3 favour mutual deposition on a wet surface by alteration of the pH. |6| Measurements of throughfall composition in a Dutch forest by van Breemen et al. |7| have shown the deposition of equal amounts of NH_4 and SO_4, 5 - 20 times larger then in rainwater. This probably reflects the large capability of a wet canopy to store pollutants and emphasizes the important role of NH_3 in deposition processes in the Netherlands. Interpretation of the throughfall measurements will be done in cooperation with the participating soil research groups.

In a later stage of the project an attempt will be made to calculate aerosol size distributions and profiles within the canopy to compare them with results obtained during some measuring campaigns. Recently some modelling results were published by Wiman and Ågren |9|, showing a strong influence of particle size on aerosol dispersion and depletion in a forest.

If we succeed in estimating the fluxes of the main gaseous pollutants to the vegetation under various conditions, we still have to relate these fluxes or the measured concentrations to the observed effects such as e.g. photosynthesis reduction. We expect that the simultaneously running cuvette experiments will provide extra information on the physiological responses of a tree. Nevertheless there still is a significant gap in our knowledge as far as processes inside a tree are concerned. The redistribution of chemical species in tree parts and the modelling of biochemical cycles and growth remain subjects for additional research.

Another problem arises when we consider the representativity of our measurements. Dutch forest stands are rather small and often surrounded by other tree or vegetation types. Near the edges of our measuring site we will probably find quite different concentration profiles. Additional measurements with a mobile tower will have to provide more detail about the horizontal variations.

On the other hand the models we intend to use are only fit for a horizontally homogeneous situation. They will not be able to resolve concentration differences within a small and inhomogeneous forest stand. Even when we would have a large forest stand with a sufficient

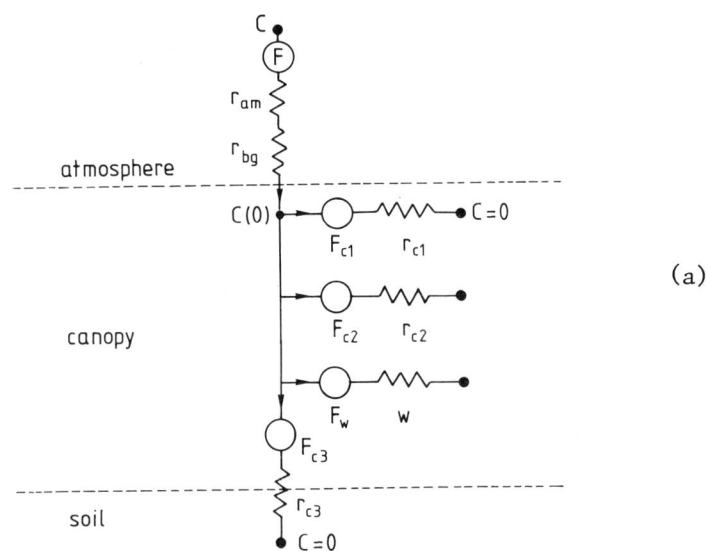

FIGURE 2. (a) A one-layer resistance analogue for pollutant uptake by a crop canopy (Grace, 1981). |2| The canopy resistance is the result of the resistances in four parallel paths: stomata (R_{c1}), cuticle and dry surfaces (R_{c2}), soil (R_{c3}) and moisture trapped on leaves (w). Fluxes F_{c1}, F_{c2}, F_{c3} and F_w travel along each path.
(b) Cross-section of a leaf, showing parallel transport ways for gaseous air pollutants near and inside a leaf. (Smith, 1981). |8|

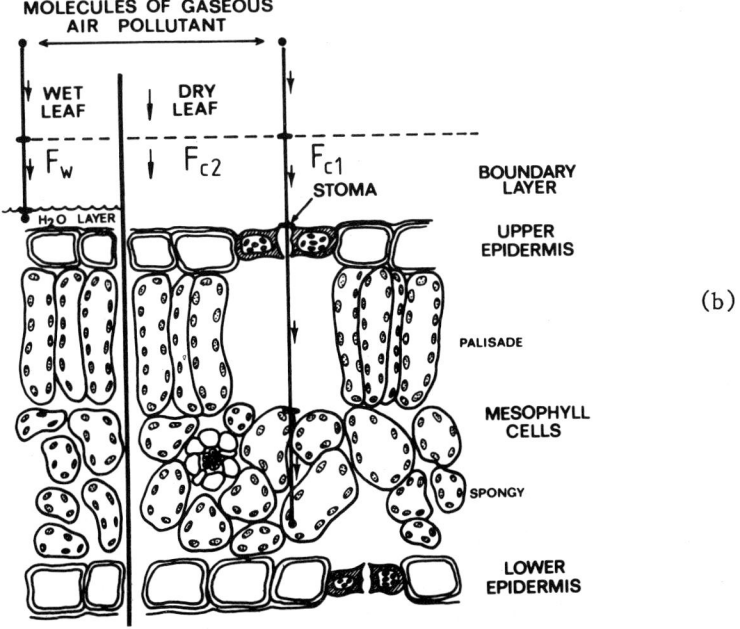

upwind fetch, we would find spatial differences in concentration profiles on the scale of the tree interspaces, due to an irregular distribution of up- and downdrafts, especially under buoyant conditions.

Aside from the spatial fluctuations there will be fast changes in time of the turbulent transport of momentum, mass and heat, which can not be detected by our 'slow' gas measuring techniques. There are even indications that in a forest a large part of the turbulent transport occurs during a few short events, due to gusts with a typical time constant of 20 to 40 seconds. (Shaw)|10| During these 'gusts' or 'sweeps' counter-gradient transport often occurs, as is evident from data presented by Denmead and Bradley |11|, who have compared fluxes, calculated from gradients, to direct eddy-correlation measurements. These sudden changes in the flow regime may be caused by large-scale eddies, generated by inhomogenities in the upwind surface roughness.

5. CONCLUSIONS

From the points stressed above it is clear that additional field research on flow characteristics as well as the dispersion of air pollutants in forests is needed badly. Especially intensive measuring campaigns during which both flux-gradient and eddy-correlation techniques are used at the right place and time are able to increase our understanding of transport processes in forests. Since the eddy-correlation method is a rather expensive technique and not always fit for continuous operation, the monitoring of pollutant concentrations under all conditions will be the first step to take.

As far as modelling is concerned we have to conclude that we are far away from a correct description of pollutant exchange in a forest. Therefore we shall, as a first approach, have to rely on the classical resistance models, often used for lower vegetation types, and make ad hoc adaptations whenever necessary.

During the next few years more information on this subject will gradually become available from this project and others of the same type in Canada, Germany, France and Switzerland.

6. REFERENCES

|1| Ministry of Housing, Physical Planning and Environment, the Netherlands: Supplementary Research Program on Acidification, June 1985.
|2| J. Grace, E.D. Ford, P.G. Jarvis (eds.): Plants and their atmospheric environment. Blackwell Scientific Publications, Oxford 1981, 111-136.
|3| H.J. van Belois: Tweedimensionaal model voor verspreiding van luchtverontreiniging boven en in een bos. (2 dim. model for air pollution diffusion in a forest). Agricultural University Wageningen, Dep. of Air Pollution, Report V-158, 1985 (in Dutch).
|4| J. Goudriaan: Crop Micrometeorology: a simulation study. Ph.D. Thesis Agricultural University of Wageningen, Pudoc Wageningen, 1977.

|5| J. Goudriaan: 'MICROWEATHER Simulation model, applied to a forest'. In: S. Halldin (ed.): Comparison of Forest Water and Energy Exchange Models, ISEM, Copenhagen, 1979.

|6| E.H. Adema, P. Heeres, J. Hulskotte: 'On the dry deposition of NH_3 SO_2 and NO_2 on wet surfaces in a small scale windtunnel'. Paper to be presented on the 7th World Clean Air Congress, Sydney, August 25-29, 1986.

|7| N. van Breemen et al.: 'Soil acidification from ammonium sulphate in forest canopy throughfall'. Nature 299, 548-550 (1982).

|8| W.H. Smith: Air pollution and forests. Springer Verlag New York, 1981.

|9| B.L.B. Wiman, G.I. Ågren: 'Aerosol depletion and deposition in forests - a model analysis'. Atmospheric Environment 19, 335-347 (1985).

|10| R.H. Shaw: 'On diffusive and dispersive fluxes in forest canopies'. In: B.A. Hutchison, B.B. Hicks (eds.): The Forest-Atmosphere Interaction. Proc. Conf. Oak Ridge, Oct. 1983, p. 407-419. Reidel Dordrecht, 1985.

|11| O.T. Denmead, E.F. Bradley: 'Flux-gradient relationship in a forest canopy'. In: B.A. Hutchison, B.B. Hicks (eds.): The Forest-Atmosphere Interaction, Proc. Conf. Oak Ridge, Oct. 1983, p. 421-442- Reidel Dordrecht, 1985.

GASDEP - GASEOUS DEPOSITION MEASUREMENTS OF SO_2, NO_x, AND O_3 TO A SPRUCE STAND: CONCEPTION, INSTRUMENTATION, AND FIRST RESULTS OF AN EXPERIMENTAL PROJECT

G. Enders, U. Teichmann
Lehrstuhl für Bioklimatologie und Angewandte Meteorologie
Universität München
Amalienstraße 52
D-8000 München 40 / FRG

ABSTRACT. Using the flux-gradient approach deposition of some gases to an adult spruce stand is determinded. Analysis of air samples simultaneously pulled down from different levels through heated Teflon tubes is done in two ways: NO_x and SO_2, respectively, from three levels are measured after an accumulation process by one monitor, so that the gradients are not affected by individual analyzer errors. Accumulation times are 8 min and 20 min, respectively. O_3 is measured on-line every 10 s by different analyzers. The same time resolution applies to meteorological readings used to determine turbulent diffusion. Performance of accumulation units and control logic, both just recently developed, and behavior of sample lines were studied in a test experiment in the 'Ebersberger Forst' near Munich, before the main investigation in a remote National Park will start. Objectives and instrumental design are reported, some results from the test study are discussed.

1. INTRODUCTION

Speaking of 'Acid Deposition' generally all three phases are understood, the gaseous as well as the fluid and solid one. While fluid and, with restriction, solid input processes have been investigated for years, there is still a considerable research deficit in gaseous deposition, at least to forest ecosystems. One of the reasons for it may be seen in the comparatively extensive and expensive instrumentation, which is required, when meteorological and chemical sensors have to be brought into the forest-atmosphere boundary layer.
    This layer has been a major point of interest in the research activities of the Institute for Bioclimatology and Applied Meteorology as a member of the Faculty of Forestry. Especially the interactions between forest and atmosphere have been studied for many years. For examples see Baumgartner (1969, 1971), Strauß (1971) and Mayer (1976, 1978).

2. OBJECTIVES

In 1984 the institute began to plan the project 'Gaseous Deposition of SO_2, NO_x, and O_3 to a Spruce Forest and Parameterisation of Deposition Velocities' (GASDEP). The objectives of this study are as follows:
- measurements of concentrations and meteorological parameters at different levels below and above the canopy;
- calculation of total fluxes (from the atmosphere to the canopy) and partial fluxes (within and below the canopy);
- calculation of daily and seasonal variations of the specific deposition velocities;
- parameterisation of the deposition velocities with respect to weather (synoptic and local scale) and stand conditions.

The experimental part has been splitted up in a test phase near Munich and a main investigation, which begins in May 1986 after having moved to the National Park 'Bayerischer Wald' on the border to Czechoslovakia.

With the begin of the main investigation air analysis will be done at six levels for SO_2 and NO_x, respectively, and three levels for O_3. So disciplines, which mainly work on the effects caused by air pollutants, can be provided with concentration profiles required. However, well to the fore is the question, where the main attack of harming gases takes place, and whether forest damages are caused by influences primarily to the assimilation elements or more indirectly to the soil. This project, of course, may only proof pathways and locations of uptake by combining micrometeorological and chemical measurements. The chemical reactions, which then follow on needles, soil, and roots have to be studied in cooperation with plant physiologists and soil scientists.

3. MEASURING SITE AND METHOD

The test phase was designed to test and improve such system components, which just recently were developed and are used for the first time. It was performed 20 km southeast of Munich in the 'Ebersberger Forst', a 90 y old spruce stand. Its area of 77 km², the plane surface, and the uniform structure make it to the most homogeneous forest in Germany ideal to conduct research in forest micrometeorology. Extensive data on this forest are given by Tajchmann (1967), Klemmer (1969), and v. Droste (1969).

At present the measuring tower at the experimental plot is 41 m tall, the trees average 32.5 m in height. From the beginning the experiment was planned as a long-term investigation with contineous measurements around the clock. At that time there were no chemical sensors on the market, which operationally could have been used for the eddy correlation method. Therefore, we decided on the gradient technique, which successfully was worked with during IBP and IHD projects of the institute to study energy and CO_2 fluxes (Strauß, 1971; Hager, 1975). This approach implies, that the flux F of a gas through a horizontal plane can be expressed by the product of the vertical exchange coefficient K and the vertical concentration gradient dc/dz: $F = K \cdot dc/dz$.

4. INSTRUMENTATION

4.1. Meteorology

To calculate the turbulent diffusion coefficient K vertical profiles of air temperature, water vapor, and horizontal wind speed are measured within and above the stand. Other readings made include shortwave and longwave radiation (upwards and downwards) in two levels, soil heat flux at three spots, atmospheric pressure, and wind direction (Fig. 1). Depending on atmospheric conditions either the aerodynamic approach or the energy-balance approach may be used to determine K.

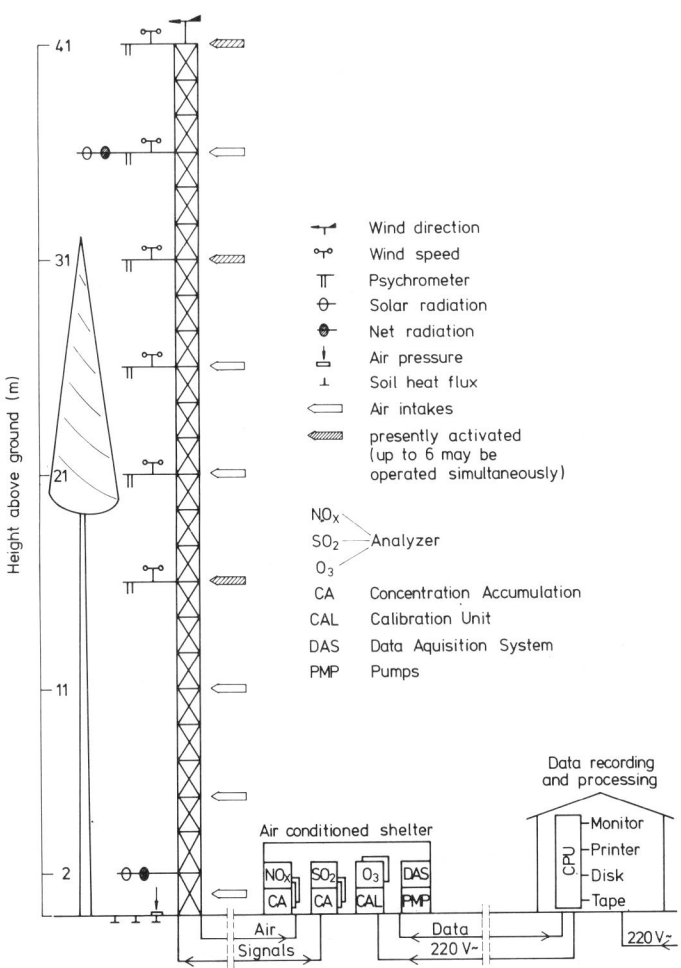

Figure 1. Instrumentation set-up at the micrometeorological station 'Ebersberger Forst' during the test experiment.

4.2. Air Sampling System

Concentration gradients of SO_2, NO_x, and O_3 in a forest atmosphere can be very low even near the canopy or the ground. That demands a high accuracy in measuring the concentrations, which at first sight could be achieved by connecting each air intake close to an analyzer and so reducing losses in the sample lines. In the open, however, the calibration of monitors to each other to the accuracy required and keeping them running under stable conditions is extremely difficult. To ensure constant operation each analyzer should have an air conditioned shelter, what may pollute temperature and wind field along the tower.

Because there was no way to meet all requirements simultaneously the following air sample system was installed: Air from different levels is contineously pulled down to a single air conditioned container at the base of the tower by a heavy duty pump with a speed of 25 m/s. The piping is composed of a modulare system, which allows a quick disassembly for inspection, cleaning or repair. Such a module is 5 m long and consists of six Teflon tubes (Ø 4.75 mm), heating wires and insulation material, all covered by a PVC protection tube. A terminal at the end contains electrical connections for the heating and Teflon fittings, which lead either to the next module or to an air intake shielded by a stainless steel cap upside down. It protects against rain and produces a slowly ascending air current inside, in which particles larger than 0.5 um are removed by gravity. Under normal operation the sampling lines are heated 20 K above air temperature to avoid condensation and reduce wall effects. For special investigations this temperature difference can be varied.

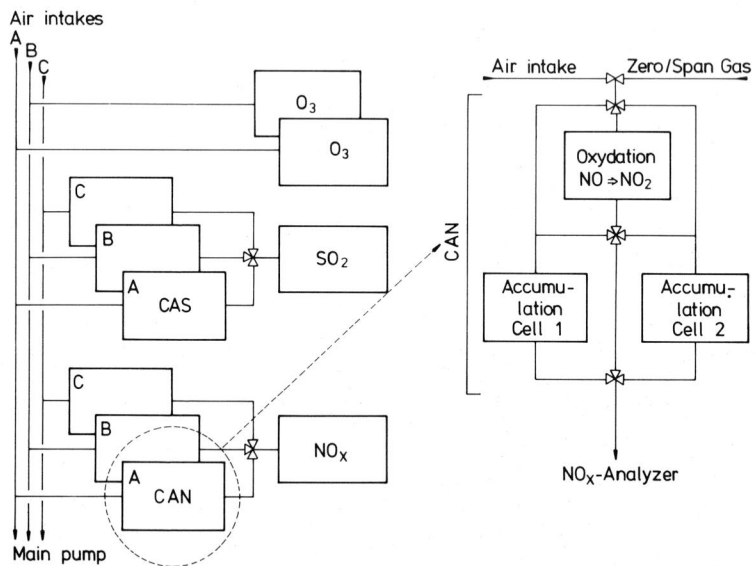

Figure 2. Schematic diagram for analysis of air samples from three levels A, B, and C using CAS and CAN accumulation devices.

Just before the level sample tubes enter the main pump individual lines exite (Figure 2) and lead either to O_3 analyzers (ML 8810, UV absorption) or to concentration accumulation units for $\underline{S}\underline{O}_2$ (CAS) and $\underline{N}\underline{O}_2$ (CAN). The prototypes of these units were developed by the Fraunhofer Institute for Atmospheric Environmental Research, manufacturing in advanced engineering is done by KONTRON Instruments under license of the Bavarian State Ministry for Environmental Protection. CAS and CAN work with the same basic measuring principle: At normal temperature SO_2 and NO_2, respectively, are trapped by specific chemical absorbers, while the air flows through. After some time of accumulation the absorbers are heated to a degree, that the gases become free again, and desorption is done. The absorption and desorption times presently used are 20'/5'30'' for SO_2 and 8'/1'55'' for NO_2. Three by three CAS and CAN are connected to a SO_2 analyzer (ML 8850, UV fluorescence) and a $NO/NO_2/NO_x$ analyzer (ML 8840, chemoluminescence), so that concentrations from three levels are measured by one analyzer. Gradients, therefore, are not falsified by different analyzer errors. To get NO_x the NO has to be oxidized before entering the accumulation cell. All CAS, CAN and analyzers have external pumps to provide the flow required.

Sampling cycles for one species simultaneously start and end to ensure gradients, which are not affected by a time lag in concentration measurements. Desorption phases, however, take place one after the other. To reduce the time gap, which otherwise would occur, each CAN and each CAS is supplied with a second accumulation cell. One of them is in the absorption mode, while the other is beeing desorbed or waiting for it (Figure 3). This 'handshaking' is controlled by an IBM personal computer, which switches the solenoid valves and takes contineously the readings of the analyzers. It also corrects the concentration

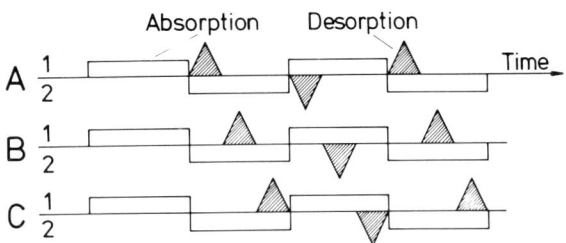

Figure 3. Timing of absorption and desorption phases, when air samples from levels A, B, and C are measured with two-way CAS or CAN units.

values resulting from a peak integration by the ratio of desorbed to adsorbed volume. This ratio has to be determined from time to time for each accumulation cell by measurements. Enrichment process and peak integration reduce the detectable limit to arround 50 ppt.

If special investigations require a higher time resolution in measuring SO_2 or NO_x, the accumulation units may be passed by. In this case, however, only one level at a time can be analyzed.

Calibration cycles are performed only on request. Zero and span gases for analyzers and accumulation units are provided by a ML 8550 calibrator. Alternatively, a complete cycle may be controlled by the IBM processor or may be performed in single steps.

4.3. Data Acquisition And Processing System

The different tasks of controlling the air sample system, scanning the analyzer peaks with a frequency of about 5 Hz, and calculating the concentrations keep the IBM computer busy. Therefore, the measurements of all other sensors is burdened to a separate data acquisition system with a KONTRON PSI 980 R/C as main unit. After multiplexing analog signals from thermometers, radiometers, heat flux plates, pressure-gage, wind vane, and O_3 analyzers are converted to digital signals by ADC's. Electronic pulses from the wind profile system are measured by counters. The concentration values are received from the IBM processor as ASCII strings via a RS232 interface. Presently, 42 channels are measured every 10 s, even the SO_2 and NO_x values to keep the file structure uniform. All data are displayed in real-time as raw data. For sensor calibration and checks on electronic components each channel may be selected individually. Then analog value (if existing), digital value and evaluated value in engineering units are shown.

When used in the operation mode the processor acts as a slave to a master computer. Planning the computer system it became clear, that the possibility of an on-the-spot data processing would be enormously helpful in detecting complex system failures. It also would allow to perceive special events and react, when modifications in measuring are recommended. The preconditions for that are given, because the station ist almost permanently manned.

Therefore, the data acquisition system was extended to a data processing system, which due to the noise and the limited space in the container is placed separately in a lodge 50 m from the tower. (The lodge also houses workrooms and simple overnight accomodations. A similar lodge will be available during the main experiment at a distance of 150m from the tower.) Pre-processor and master are connected by a fiber optic. The master station consists of a PSI 980 Q computer with 256 KB memory, graphics capable printer and monitor, a 20 MB hard disk, a floppy disk drive with a capacity of 616 KB, and a magnetic tape unit. The raw data received from the satellite processor supplied with date and time are stored in their original form on the hard disk, which may hold more than 14 complete days. Simultaneously, the data can be displayed in real time after a conversion into engineering units by scale functions stored in a data description file. This file contains for each channel the information needed to manage the measurement, e.g. sensor type, status, location, scale function, validity range, etc. It is also used, when retrieval programs call for data already stored. Further data exploitation is made possible by statistical software packages and programs developed by ourselves. Languages available are FORTRAN, BASIC and PASCAL. Copies of the data files are made to tapes to free space on the hard disk and to allow extensive calculations on a IBM 3081 located in Munich, when more memory space and CPU-time is needed.

5. RESULTS AND DISCUSSION

5.1. Comparison Of O_3 Measurements

Despite of the high speed of the air current inside the level sampling lines and their heating, concentration losses due to condensation and wall effects may not be excluded. According to the chemical reaction scheme the O_3 should be affected most among the species under investigation. Therefore, a comparison study of O_3 measurements was performed in September 1985. Analyzer 1 was moved up to a height of 31 m and connected to the air intake by a short unheated Teflon manifold (Figure 4). Analyzer 2 remained in the container and was operated under normal conditions. The total length of the heated line was 45 m, its volume 0.8 l, the delay time for an air sample between intake and analyzer 2 1.8 s. Until the begin of the comparison experiment the system was run approximately 2 800 h, so that about 4 500 m³ air were pulled through this level sampling line.

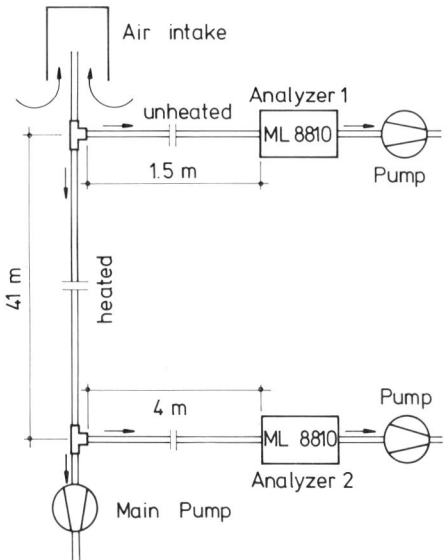

Figure 4. Arrangement of O_3 analyzers to determine ozone losses.

Both analyzers were calibrated in the container at a temperature of 22° C and a air pressure of 958 hPa. However, due to the pressure drop in the sampling line analyzer 2 was operated later on at a pressure of around 710 hPa measured just before the photometer. That reduced the extinction in the analyzer, which operates according to Beer's law. The extinction coefficient, however, was not set to this pressure condition, because the manometer available had an error of ± 3 %. Therefore, concentrations given by analyzer 2 are too low. Analyzer 1 operated nearly under calibration conditions, because the air pressure remained almost constant and the air temperature at 31 m varied only between 15° C and 25° C causing nearly no effect.

Figure 5 gives the results for 34 hours of the experiment. Concentrations shown are 10 min averages computed from the original 10 s data. The main difference between analyzer 1 and 2 was 6 ppb with a standard deviation of 1 ppb. The maximum difference was 9 ppb, the minimum 3 ppb. Correlations made did not show a significant influence of any meteorological parameter to the variation. On the other hand, the differences can not easily be explained by tubing losses, because the unavoidable error in calibrating the two analyzers to each other and the error made in using a too small extinction coefficient for analyzer 2 sum up to a total error of a comparable magnitude. This shows the need to repeat this study with a micromanometer (tolerance < 1 ‰) recently purchased for this purpose. Baumbach und Käß (1985) used for a similar study a single analyzer, which alternatively was switched to a calibrator either directly or via long (up to 27 m, Ø 13 mm) heated Teflon tubes. The O_3 losses determined were maximal of the same size as the differences between different analyzers of the same type (Dasibi 1008 PC).

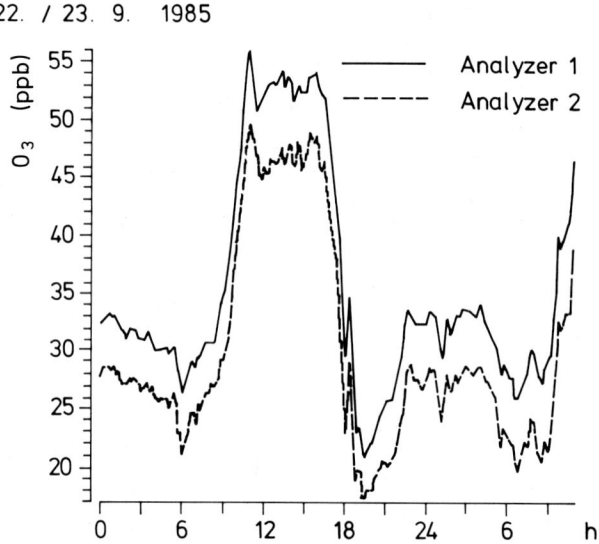

Figure 5. O_3 concentration patterns (10 min averages) in 31 m height measured by two analyzers during 34 h of the comparison study (analyzer 1 at 31 m, analyzer 2 at the ground).

5.2 Comparison Of NO_x Measurements

During October 1985 a study similar to the O_3 comparison was made with respect to NO_x. An interpretation of the results is even more presumptuous, because the analyzer at the tower had to be operated under cold and foggy weather conditions quite different to the calibration conditions. Again the pressure at the end of the long tube could not have been determined to the accuracy required. Therefore, only the pure figures are reported.

In contradiction to the behaviour of O_3 the differences in NO_X were the higher, the higher the concentration was. While the concentration measured by analyzer 1 at 31 m decreased from a maximum of 76 ppb to 7 ppb, the differences to analyzer 2 at the ground run down from 36 ppb to -0.4 (!) ppb. Because the concentration was less than 25 ppb during 26 of 32 hours of observation, the mean difference is only 5 ppb with a standard deviation of 6 ppb. All numbers given are again 10 min averaged.

As mentioned before it was planned to repeat the comparison studies with a more accurate manometer and extend them also to SO_2. To have then well defined conditions inside the piping system, the tube modules were disassembled, cleaned and mounted again. Due to the early and severe winter invasion, which followed, the comparison experiments had to be postponed. Before and after cleaning one of the modules was flushed with Na_2CO_3 to get a first qualitative idea of the contamination and the efficiency of cleaning. NO_3 was reduced from 0.09 to 0.05, $SO4$ from 2.17 to 0.11, $NH4$ from 2.70 to 0.06, all numbers given in mg/l. This cleaning and analysis process shall be repeated during the main investigation on a regular basis.

5.3. NO_X Flux And Deposition Velocity

The main question to be answered after the test experiment is, of course, how the system performs as a whole. Flux calculations, for which it was designed, could not be done, before the performance of each individual hardware and software component and their co-operation were sufficiently checked out. The main problems were: repeated destruction of the heating wires of the accumulation cells; short circuits in main pump and tube heating power supply; incorrect data transmission between the IBM computer and the pre-processor due to a wrong timing, which occured more or less randomly; actuation of solenoid valves at a wrong time or in a wrong order; complete data transmission failures between slave and master unit. By now most of the problems are overcome.

There are reliable data sets, by which preliminary flux calculations could be made to see how they fit to results obtained somewhere else. Figure 6 shows the patterns of mean NO_X concentrations at 41 m and 26 m for the afternoon of August 26, 1985. Averaging time is 10 min. (If accumulation is used mean concentrations given for periods other than absorption time are weighted means). The maxima occured around 17h, when the wind came from $300°$ pointing to the end of a highway with frequent traffic jams. Wind speed at 41 m averaged to 2-3 m/s this afternoon. O_3 concentration at 41 m was comparably low due to overcast sky. After some rain in the morning the canopy was still wet. Air temperature at 41 m ranged between 9°C and 11°C at stable conditions in the atmosphere below.

In figure 7 the variation in the turbulent diffusion coefficient K, the NO_X flux F, and the NO_X deposition velocity v_g are shown. Positive

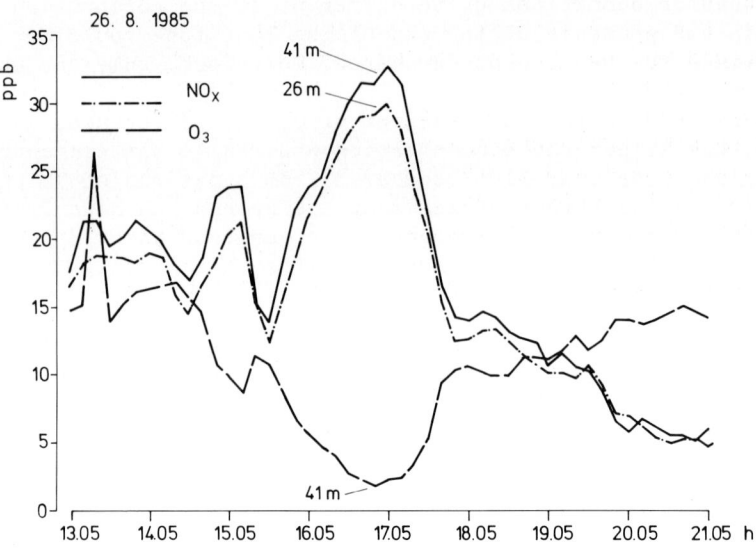

Figure 6. 10 min averaged NO_x and O_3 concentrations on August 26, 1985 from 13:00 until 21:00.

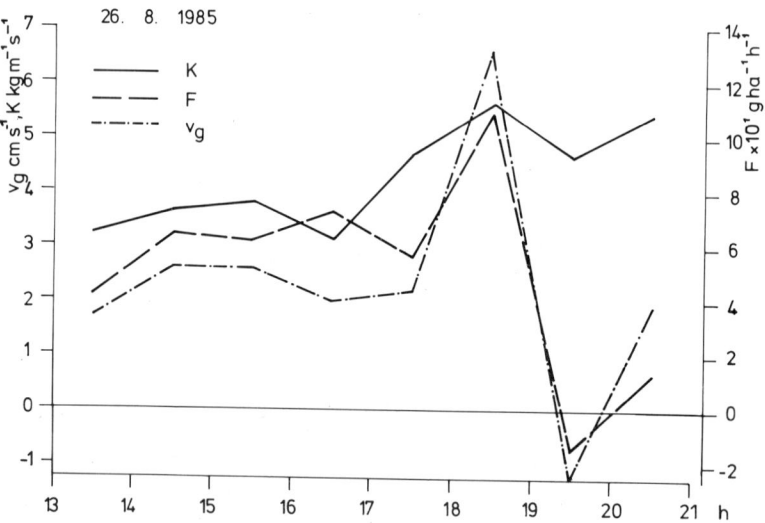

Figure 7. 60 min averaged variations in the turbulent diffusion coefficient K, the NO_x flux F, and the NO_x deposition velocity v_g above the canopy layer on August 26, 1985 from 13:00 until 21:00. Positive signs indicate downward fluxes.

signs for F and v_g indicate downward fluxes. Values given are averaged by one hour. K reached its maximum between 18:00 and 19:00 causing the maxima of F and v_g. Due to inverse NO_x gradients between 19:00 and 20:00 even upward fluxes result from the gradient approach, deposition velocities become negative. That may be true, because small negative gradients could be seen again and again during the whole test experiment. On the other hand, Denmead and Bradley (1985) made the existence of countergradient fluxes of sensible heat, water vapor, and CO_2 in and below the canopy evident using simultaneously the eddy correlation method. However, "in the upper parts of the needle space there was almost universal correspondence between the directions of the fluxes and the signs of the appropriate gradients." May be, that having taken measuring heights of 31 m or 36 m instead of 26 m for the second NO_x intake would have led to an other result for the flux above the canopy layer, but this was not really important for the main conclusion we like to draw: The deposition velocities calculated for this special day seem to fit the range of results cited in the literature, the instrumental set-up can be used to work with the gradient approach.

Therefore, confidently we begin now with the main investigation in the National Park 'Bayerischer Wald'.

6. ACKNOWLEDGEMENTS

This paper was prepared from work granted by the Federal Ministry of Research and Technology under contract number 03-7339-0. Data processing on the IBM 3081 is funded by the Bavarian State Ministry of Food, Agriculture and Forestry, which also provids the forest sites. The authors would like to thank the scientific and technical staff of the institute for their support, particularly Prof. Dr. A. Baumgartner, who headed the project. We are grateful also to Mr. W. Brensing from Kontron Instruments for providing technical advice, Mrs. C. Mooslechner for drawing the figures, and Mrs. H. Danÿ for typing the paper.

7. REFERENCES

Baumgartner, A.: 1969. Meteorological Approach to the Exchange of CO_2 between the Atmosphere and Vegetation, particularly to Forest Stands. Photosynthetica 3:127-149.

Baumgartner, A.: 1971. Wald als Austauschfaktor in der Grenzschicht Erde/Atmosphäre. Forstw. Cbl. 90:174-182.

Baumbach, G. and M. Käß: 1985. Ermittlung von vertikalen Schadstoffkonzentrationen in Waldbeständen in Baden-Württemberg - Meßtechnik und erste Ergebnisse. Staub - Reinhalt. Luft 45:274-278.

Denmead, O.T. and E.F. Bradley: 1985. Flux-Gradient Relationships in a Forest Canopy. pp. 421-442 in 'The Forest-Atmosphere Interaction', B.A. Hutchison and B.B. Hicks (eds.). D. Reidel Publishing Company.

Droste-Hülshoff, B. von: 1969. Struktur und Biomasse eines Fichtenbestandes auf Grund einer Dimensionsanalyse an oberirdischen Baumorganen. Diss. Staatsw. Fak. Univ. München. 219 pp.

Hager, H.: 1975. Kohlendioxydkonzentrationen, -Flüsse und -Bilanzen in einem Fichtenhochwald. Wiss. Mitt. Meteorol. Inst. Univ. München No. 26. 182 pp.

Klemmer, L.: 1969. Die Periodik des Radialzuwachses in einem Fichtenwald und deren meteorologische Steuerung. Wiss. Mitt. Meteorol. Inst. Univ. München No. 17. 85 pp.

Mayer, H.: 1976. Die Windverhältnisse in und über einem Fichtenwald. Forstw. Cbl. 95:333-345.

Mayer, H.: 1978. Kenngrößen des vertikalen Windprofils in und über einem Fichtenwald. Agric. Meteorol. 19:275-293.

Strauß, R.: 1971. Energiebilanz und Verdunstung eines Fichtenwaldes im Jahre 1969. Wiss. Mitt. Meteorol. Inst. Univ. München No. 22. 66 pp.

Tajchman, S.: 1967. Energie- und Wasserhaushalt verschiedener Pflanzenbestände bei München. Wiss. Mitt. Meteorol. Inst. Univ. München No. 12. 95 pp.

Comparison of SO_2-concentrations and immission-rates with bulk precipitation data.
- Results from the Hessian research program "Forest Stress by Air Pollution-WdI" -

K. Hanewald, Hessische Landesanstalt für Umwelt, Wiesbaden

Abstract

The following report gives a brief description of the subproject "immission registration", a part of the Hessian measuring program "Forest stress by air pollution-WdI". Some of the data from Witzenhausen will be discussed. These data include the SO_2-concentration, the SO_2-immission rate, the amount of precipitation as well as the sulfate and nitrate concentrations in bulk deposition. A statistical interpretation indicates that an increase in particulate nitrate and sulfate, when compared with wet deposition, is found during winter periods with often occurring stagnating air masses. During summer, the greatest amount of sulfate and nitrate input is found in form of wet deposition.

1. Brief introduction to the research program

The Hessian Environmental Agency (HLfU) is operating measuring stations in Königstein, Grebenau and Witzenhausen since May 1983 to ascertain the immission input into the forest ecosystems.

Two further stations, near Frankenberg and Biebergemünd/Spessart were started up in Oct./Nov. 1985 to extend the measuring network in Hesse. The station to be started in Spring 1986 near Fürth/Odenwald, will be the sixth and final station. The maximum air distance between stations is 60 km. The locations are shown in Figure 1.

Measurements at the first locations near Königstein, Grebenau and Witzenhausen consist of the quantitative determination of gaseous, liquid and particulate pollution input at several levels in 60 - 80 years old spruce stands as well as in nearby located open field stations.

The principle is shown in Figure 2.

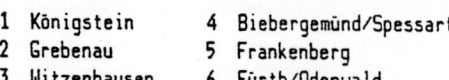

FIGURE 1 : Measuring sites 'Forest Stress by Air Pollution'

1 Königstein   4 Biebergemünd/Spessart
2 Grebenau   5 Frankenberg
3 Witzenhausen   6 Fürth/Odenwald

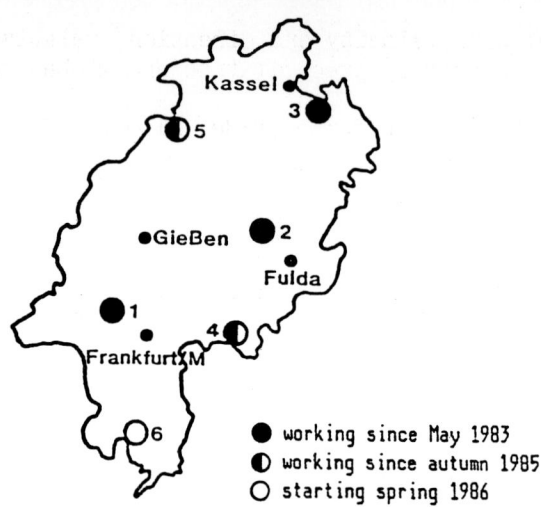

● working since May 1983
◐ working since autumn 1985
○ starting spring 1986

LEVEL 1 :
5 - 7 m above tree canopy, allows the interpretation of large scale transport of air pollutants

LEVEL 2 :
in the tree canopy, determines the amount of pollution interacting with the assimilation organs (needels)

OPEN FIELD
continuous immission measurements

LEVEL 3 :
registration of data about filtering and/or leaching effects

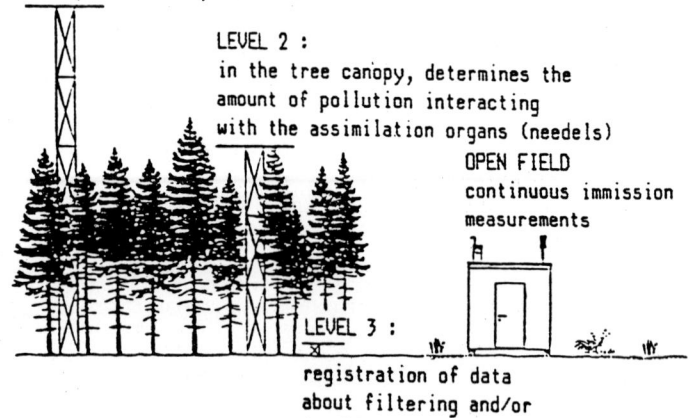

FIGURE 2

The new stations near Biebergemünd, Frankenberg and Fürth will not have towers but open field gages. The three existing towers are sufficient for special investigations in spruce stands, for example gradient measurements.

The following parameters are determined:

1. Meteorology: wind velocity and direction, relative humidity, temperature, amount of precipitation and global radiation

2. Gaseous components: sulfur dioxide, nitrogen dioxide, nitrogen monoxide and ozone

3. Contents of bulk deposition and dust: Pb, Mn, Cd, Al, Cu, Cr, Mg, Ca, Na, K, SO_4^{2-}, Cl^- and NO_3^-

4. Immission rates (IRMA): sulfate, nitrate, fluoride.

A more detailed description of the research program is given in the first interim report (1).

2. Methods and chosen periods

The second interim report (2) gives a detailed description of results from the continuous gaseous immission measurements.

A comparison of precipitation amounts, sulfate and nitrate bulk deposition, the SO_2 immission rates as well as the results of continuous SO_2 immission measurements is given, whereby this first attempt at correlation is restricted to the Witzenhausen WdI station in northern Hesse.

Bulk precipitiation sampling was done with the "Münden 100" precipitation gage (3) over 14 day periods. Nitrate and sulfate were determined by ion chromatography (4) at the Institute for Meteorology and Geophysics, University of Frankfurt. The SO_2 immission rate determination went also over 14 day periods in accordance with the VDI regulation 3794, paper 1 (5). The continuous SO_2 measurements were carried out in accordance with VDI regulation 2451, paper 4 (6). The half hourly SO_2 concentration averages were converted into 14 day averages.

Figures 3a - c and 4 a + b illustrate the 14 day averages for the Witzenhausen station between 01 Nov. 1983 and 31 March 1985. With the exception of the sulfur dioxide concentrations ($\mu g/m^3$) which were measured continuously at the open field station, all results for the amount of precipitation, sulfate-S, nitrate-N and the SO_2 immission rates from the tree canopy and ground level are compiled into graphic representations.

Pollution input is given as mg/m^2/d and the amount of precipitation as mm/d.

The average quotients between the canopy and ground levels for the entire period will be seen in the legend at the right margins, whereby the canopy average is defined as 1.

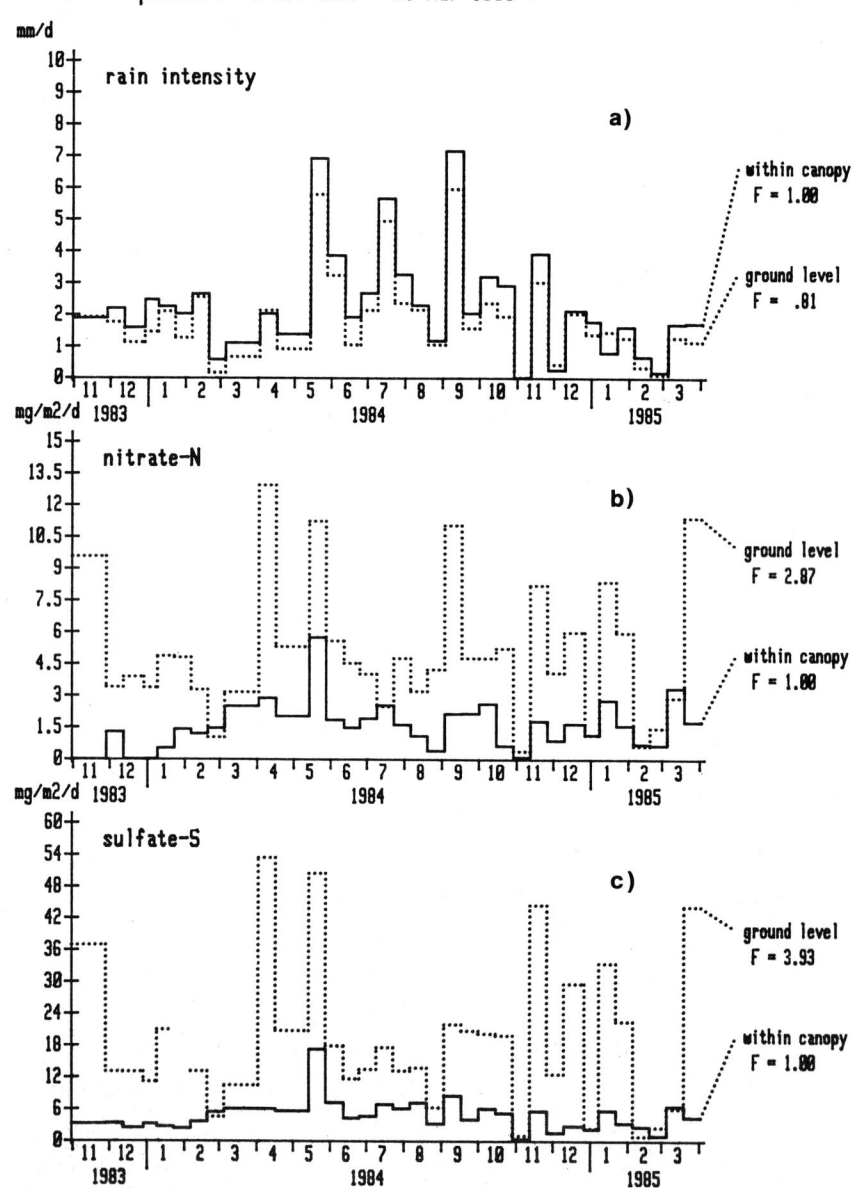

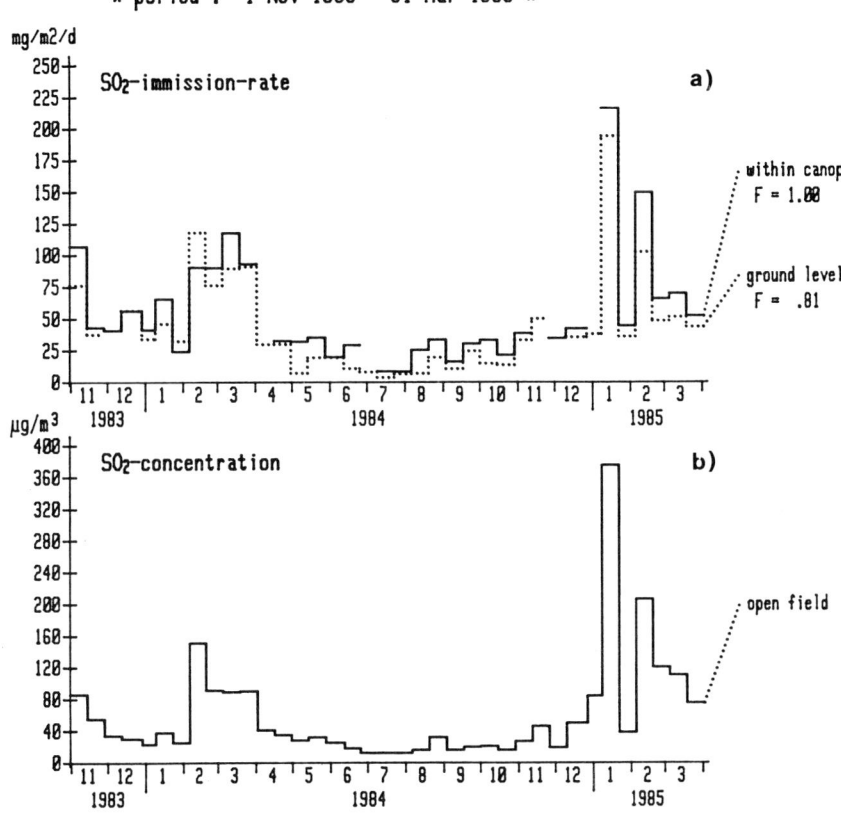

FIGURE 4 * HLfU * WdI * Witzenhausen * 14 day means *
* period : 1 Nov 1983 - 31 Mar 1985 *

The greater than normal amount of precipitation in summer 1984 results in summer maxima for NO_3-N and SO_4-S. Their input during periods of reduced precipitation amounts (here especially in Dec. 1984 - March 1985) attains, in individual values, the same range as found in summer.

Summer minima, as expected, are found in the immission rates and SO_2 concentrations. Smog periods in January 1985 account for extraordinary high concentrations and deposition values per unit area. A detailed analysis of the smog periods has been published in the SMOG report, 1985 (7).

3. Correlation

Correlation analyses were undertaken in order to understand better the data collective. A nonparametric rank correlation coefficient

TABLE 1 'WdI' station Witzenhausen: rank correlation coefficient according to Kendall
periods: 1 Nov 1983 - 14 Apr 1984 and 26 Dez 1984 - 31 Mar 1985

```
        1.    2.    3.    4.    5.    6.    7.    8.    9.   10.   11.   12.   13.   14.   15.
1.     ---  +,21  -,47  -,38  +,37  -,15  +,44  -,21  +,72  +,30  +,77  +,37  +,32  +,17  +,44
2.      19   ---  -,24  -,11  +,05  +,34  +,08  +,40  +,26  +,69  +,17  +,71  -,07  -,17  +,05
3.      19    19   ---  +,73  -,05  +,34  -,18  +,27  -,63  -,25  -,58  -,34  -,39  -,37  -,45
4.      19    19    19   ---  +,04  +,50  -,10  +,42  -,41  -,11  -,47  -,23  -,19  -,20  -,20
5.      19    19    19    19   ---  +,18  +,58  +,09  +,42  +,13  +,33  +,14  +,28  +,20  +,26
6.      18    18    18    18    18   ---  +,05  +,92  -,10  +,39  -,23  +,29  -,19  -,17  -,21
7.      19    19    19    19    19    18   ---  -,01  +,31  +,07  +,60  +,13  +,20  +,08  +,22
8.      19    19    19    19    19    18    19   ---  -,11  +,42  -,23  +,36  -,25  -,24  -,25
9.      19    19    19    19    19    18    19    19   ---  +,40  +,64  +,44  +,47  +,34  +,53
10.     18    18    18    18    18    18    18    18    18   ---  +,20  +,90  -,01  -,07  +,05
11.     19    19    19    19    19    18    19    19    19    18   ---  +,29  +,30  +,19  +,44
12.     19    19    19    19    19    18    19    19    19    18    19   ---  +,05  -,05  +,13
13.     17    17    17    17    17    16    17    17    17    16    17    17   ---  +,84  +,61
14.     18    18    18    18    18    17    18    18    18    17    18    18    16   ---  +,61
15.     19    19    19    19    19    18    19    19    19    18    19    19    17    18   ---
```

TABLE 2 'WdI' station Witzenhausen: rank correlation coefficient according to Kendall
period: 15 Apr 1984 - 25 Dec 1984

```
        1.    2.    3.    4.    5.    6.    7.    8.    9.   10.   11.   12.   13.   14.   15.
1.     ---  +,41  -,29  -,29  -,01  +,00  -,03  -,03  +,63  +,43  +,50  +,43  +,05  -,11  +,03
2.      18   ---  -,30  -,36  -,27  +,30  +,09  +,30  +,36  +,87  +,57  +,75  +,32  +,30  +,34
3.      18    18   ---  +,89  +,67  +,37  +,42  +,38  -,39  -,33  -,31  -,47  -,50  -,19  -,19
4.      18    18    18   ---  +,64  +,34  +,45  +,32  -,39  -,36  -,28  -,53  -,42  -,21  -,17
5.      18    18    18    18   ---  +,34  +,42  +,32  -,05  -,25  -,14  -,36  -,42  -,33  -,18
6.      18    18    18    18    18   ---  +,47  +,65  -,02  +,30  +,17  +,07  +,07  +,23  +,31
7.      18    18    18    18    18    18   ---  +,32  -,02  +,04  +,27  -,09  -,10  -,13  -,03
8.      18    18    18    18    18    18    18   ---  -,06  +,24  +,13  +,14  +,00  +,29  +,39
9.      18    18    18    18    18    18    18    18   ---  +,46  +,55  +,50  +,18  -,04  +,11
10.     18    18    18    18    18    18    18    18    18   ---  +,59  +,69  +,33  +,27  +,32
11.     18    18    18    18    18    18    18    18    18    18   ---  +,47  +,27  +,08  +,18
12.     18    18    18    18    18    18    18    18    18    18    18   ---  +,24  +,20  +,27
13.     16    16    16    16    16    16    16    16    16    16    16    16   ---  +,61  +,66
14.     17    17    17    17    17    17    17    17    17    17    17    17    15   ---  +,69
15.     18    18    18    18    18    18    18    18    18    18    18    18    16    17   ---
```

TABLE 3 Parameter list

Bulk deposition:

1 Conductivity canopy (uS/cm)
2 Conductivity ground level (uS/cm)
3 Precipitation canopy (mm/d)
4 Precipitation ground level (mm/d)
5 Sulfate-S canopy (mg/m2/d)
6 Sulfate-S ground level (mg/m2/d)
7 Nitrate-N canopy (mg/m2/d)
8 Nitrate-N ground level (mg/m2/d)
9 Sulfate-S canopy (mg/l)
10 Sulfate-S ground level (mg/l)
11 Nitrate-N canopy (mg/l)
12 Nitrate-N ground level (mg/l)

Immission rates:

13 SO2 canopy (mg/m2/d)
14 SO2 ground level (mg/m2/d)

Continuous measurement:

15 SO2 (ug/m3)

according to Kendall (8,9,10) was calculated considering the ties because a normal distribution was not expected; and it was attempted to reduce by this the effect of outliers on the correlation coefficients.

Furthermore, the data were divided into two time periods in order to determine seasonal changes.

The summer period is defined as the low SO_2 immission period (see Fig. 4b) between 15 April 1984 and 25 December 1984. The remaining time periods (01 Nov. 1984 - 14 April 1984 and 26 Dec. 1984 - 31 March 1985) with occasionally increased SO_2 concentrations are defined as the winter period.

Tables 1 and 2 contain the calculation results for both periods.

The rank correlation coefficients are shown in the upper right triangle and the number of pairs per individual correlation are grouped in the left lower triangle. The meaning of the marginal numbers is given in the parameter list (Table 3). It must be mentioned that the conductivity of the bulk deposition samples as well as the SO_4-S and the NO_3-N concentrations were used as additional parameters.

Table 4 gives several significance levels with their corresponding correlation coefficients:

Table 4

level of significance (%) for n = 17	correlation coefficient
98.4	$\pm$ 0.42
99.0	$\pm$ 0.44
99.3	$\pm$ 0.47
99.6	$\pm$ 0.50

Correlation coefficients above $\pm$ 0.42 are discussed in the following considerations.

The examination of Tables 1 and 2 indicates several trivial correlations which occur in both observation periods. Examples are:

        3 - 4 (amount of precipitation in tree canopy and at ground level)

    1 - 9, 1 - 11 (Conductivities and SO_4-S respectively NO_3-N

and 2 -10, 2 - 12 concentrations)

    13 -15, 14 - 15 (immission rates and SO_2 concentrations)

These trivial correlations are not completely meaningless because they are a welcome help as feasibility controls for the newly completed computing program.

Further examination indicates close relationships between nitrate and sulfate concentrations respective their deposition rates in both periods: 5 - 7, 6 - 8, 9 - 11 and 10 - 12. This fact is interpreted by common sources for sulfate and nitrate immissions. The principal differences between the periods are:

Summer period

The only definite relationships which are found only in summer as positive correlations are:

> 3 - 5 (amount of precipitation in the canopy and
and 3 - 7 sulfate, respectively nitrate deposition)

Winter period

There are several additional relationships observable in this time period which either do not appear in summer or have a very low level of significance.

For example, the amount of precipitation in the canopy has a definite negative correlation with the corresponding sulfate and nitrate concentrations:

> 3 - 9 and 3 - 11

An obvious interpretation is that dry deposition has been diluted by rain and snow.

Further correlations are found between conductivity (with the trivially associated sulfate and nitrate concentrations) in the canopy and the SO_2 immission concentrations:

> 1 - 15, 9 - 15 and 11 - 15

Furthermore, the sulfate concentration in the canopy correlates with the corresponding SO_2 immission rates (9 - 13). A possible explanation is that the increased sulfate and nitrate is associated to weather conditions with high SO_2 concentrations. Since both winter periods are characterized by stagnating air masses and simultaneous advection of air masses with high SO_2 concentrations from easterly directions (see also (2, 7)), it can be assumed that the amount of particulate nitrate and sulfate deposition in bulk deposition is much greater in winter than in summer, and that most of this particulate deposition originates in eastern countries.

This fact is of importance for the sudden increase in acidity that occurs in Spring after the melting of snow.

4. Conclusions

This first attempt to use statistical methods and interpretation to reduce the amount of data is not to be seen as a complete analysis of an immission complex.

A test was undertaken with a reasonable data base to see if the applied techniques could identify the correlations between the test series, and if this method may be applied to reduce the amount of data. This is especially important for the WdI program with its very diversified parameters (see the compilation in Section 1) because of the possible loss of general view in the course of the years.

5. Acknowledgements

I am grateful to the Institute of Meteorology and Geophysics, University of Frankfurt for the precise nitrate and sulfate analyses. I would like to thank Mr. W. Wunderlich for the preparation of the statistics programs and Mr. A. Siegmund for the compilation and representation of the data base. Last not least, I would like to thank the Hessian Forestry Research Station for the stimulating discussions and close and cooperative assistance.

6. References

1 Waldbelastungen durch Immissionen, 1. Zwischenbericht, Hrsg. Hess. Landesanstalt für Umwelt, Wiesbaden (1984)

2 Waldbelastungen durch Immissionen, 2. Zwischenbericht, Hrsg. Hess. Landesanstalt für Umwelt, Wiesbaden (1985)

3 Brechtel, H.M.; Hammes, W.: Aufstellung und Betreuung des Niederschlagssammlers "Münden", Hersg. Hess. Forstliche Versuchsanstalt, Hann.Münden (1984)

4 Georgii, H.W., Rohbock, E., Schmitt, G., Untersuchung des atmosphärischen Schadstoffeintrags in Waldgebieten in der Bundesrepublik Deutschland, Hersg. Universitätsinstitut für Meteorologie und Geophysik, Frankfurt (1984)

5 VDI-Richtlinie 3794, Blatt 1, VDI-Verlag, Düsseldorf (1982)

6 VDI-Richtlinie 2451, Blatt 4, VDI-Verlag, Düsseldorf (1968)

7 Erfahrungsbericht SMOG '85, Hersg. Hessischer Minister für Arbeit, Umwelt und Soziales, Wiesbaden (1985)

8 Sachs, L., Angewandte Statistik, Springer-Verlag Berlin, Heidelberg, New York, Tokyo (1984)

9 Lienert, G.A., Verteilungsfreie Methoden in der Biostatistik, Verlag Anton Hain, Meisenheim am Glan (1978)

10 Stange, K.; Henning, H.-J., Formeln und Tabellen der mathematischen Statistik, Springer-Verlag Berlin, Heidelberg, New York (1966)

11 Georgii, H.-W., Perseke, C., Rohbock, E., Feststellung der Deposition von sauren und langzeitwirksamen Luftverunreinigungen aus Belastungsgebieten, Hrsg. Universitätsinstitut für Meteorologie und Geophysik, Frankfurt (1982)

WET AND DRY DEPOSITION OF ATMOSPHERIC TRACE ELEMENTS IN FOREST AREAS

Stefan Grosch
Institute of Meteorology and Geophysics
Feldbergstraße 47, University of Frankfurt/M
6000 - Frankfurt am Main
Federal Republic of Germany

ABSTRACT. Wet and dry deposition rates of the elements SO_4^-, NO_3^-, Cl^-, Pb, Mn, Fe and Cd estimated for the period of April 1983 to April 1985 at 5 different measuring sites, most of them located in the state of Hesse, are presented.
In accordance to earlier investigations, wet deposition is found to be the dominant sink for the elements SO_4^-, NO_3^-, Cl^-, Pb and Cd. Mn and Fe are affected by dry deposition to a large extent. No pronounced differences of total deposition have been found, comparing three sampling sites in forest areas with the other two sites within the sampling network.
The importance of "episodes" of wet deposition is shown at one station for SO_4^-. Up to 40 % of the annual wet deposition of SO_4^- is associated with only 10 % of the annual rainfall-events.
Dry deposition processes that take place in the canopy of forest stands lead to an enhancement of element concentrations in the throughfall. The ratio of different elements in precipitation below and above the canpoy suggests a functional dependence on the mass median diameter of the corresponding aerosol particles.

1. INTRODUCTION

Deposition measurements, leading to useful data about occuring element fluxes, are helpful to characterize different areas according to their individual conditions of stress. The knowledge of these fluxes is especially important as basic information for those scientists who try to evaluate possible relations between observed forest damages and atmospheric pollution.
The still increasing forest decline throughout the last few years is not restricted to coniferes, which has to be referred as the more sensitve species, but is documented now also for various decidious trees like beach or oak (GSF, 1985).
In principle we have to differentiate between wet- and dry deposition as two possible pathways of atmospheric deposition. With respect to dry deposition furthermore we have to separate between gases and particulate

matter. The term "wet deposition" is used in general to describe the deposition caused by rain.
Instead of that, fog which is also "wet deposited" and of great interest as a potential cause for foliar damages of trees, usually is discussed as a separate pathway of deposition.
The routine investigations presented in this paper are restricted to measurements of wet- and dry deposition rates, the latter obtained by using the sampling method after Bergerhoff (Georgii et al., 1984).
In addition some analyses of samples of the bulk deposition, collected at different levels at the investigated forest stations are reported.
These investigations are carried out in cooperation with several official agencies and scientific institutes within a research program entitled "Waldbelastungen durch Immissionen", translated as "forest stress caused by immissions" (Waldbelastungen durch Immissionen, 1984; 1985)

2. MATERIALS AND METHODS

At a sampling network of five stations (fig. 1) the wet- and dry deposition is collected separately by means of a suitable wet and dry deposition-sampler. Detailed information about this sampler is reported elsewhere (Rohbock and Georgii, 1982).

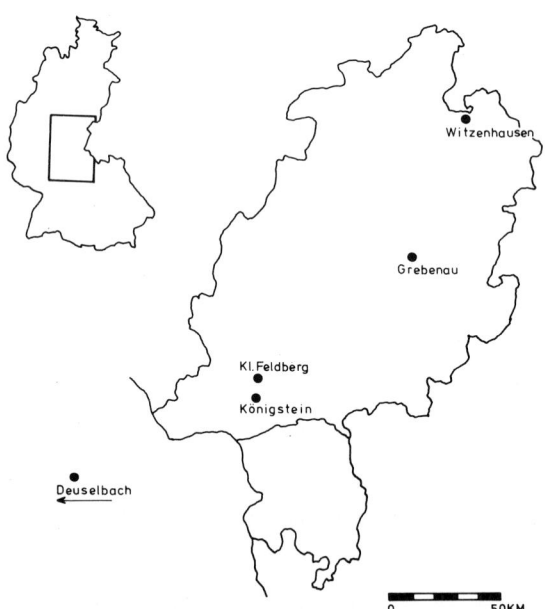

fig. 1: Geographical location of sampling sites.

The precipitation samples are collected on a daily basis. To obtain a measure for the dry deposition, a standardized method is applied using a glas vessel as sampling system (VDI, 1972). The sampling period is 14 days.
The examined sampling sites, three of them located in spruce stands (Königstein, Grebenau, Witzenhausen) has to be characterized as areas of small influence by local pollutant-sources.
Bulk deposition samples are taken below and above the canopy (on top of a tower) at all three forest sites. The sampling period here is also 14 days.
For further detailed investigastion of the throughfall, nine "wet-only samplers" has been installed at the station Königstein, in order to prevent the contamination of precipitation samples as far as possible. A sketch of the used sampler is shown in fig. 2. Samples are taken on a daily basis.

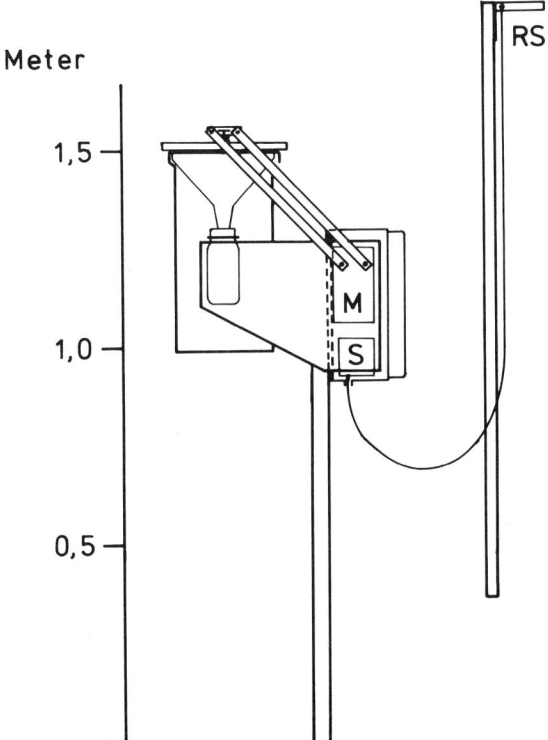

fig. 2: Wet-only sampler. RS: rain-sensor, S: electronical controlling
   M: motor

Precipitation samples were analysed for conductivity and pH at first. After appropiate treatment according to the analytical method used, all samples were analysed for SO_4^-, NO_3^- and Cl^- by means of ion chromatography and Pb, Mn, Fe, Cd, Al, Cu, Cr, Na, K, Ca and Mg by means of atomic absorption spectroscopy.

Detailed information about sample treatment is published elsewhere
(Georgii et al., 1982). The investigation started in autumn 1982. The
presented results of mean deposition rates are related to the period of
April 1983 to April 1985. All data of SO_4^- and NO_3^- are calculated as S
and N respectively.

3. RESULTS

3.1 Deposition

The wet deposition as product of concentration and total amount of
rainfall, gives a measure of element masses deposited per unit of time
and area. Fig. 3 and 4 shows the sum of wet- and dry deposition, the
latter being represented as dark part of the bars. From this it is
possible to differentiate the elements with respect to varying importance
of the respective deposition pathways.

Enhanced depositions in spite of low trace element concentrations in
precipitation can occur due to high rainfall amount as being the dominant
influence regulating the wet deposition. This is clearly realized by
comparing the two sampling sites Königstein, which shows an annual
rainfall amount of about 780 mm and Witzenhausen (about 670 mm/a). In
spite of higher concentrations in precipitation found at Witzenhausen
(except of Fe), the amounts of wet deposition at Königstein are in the
same range and sometimes even higher (for NO_3^-, Cl^- and Pb).

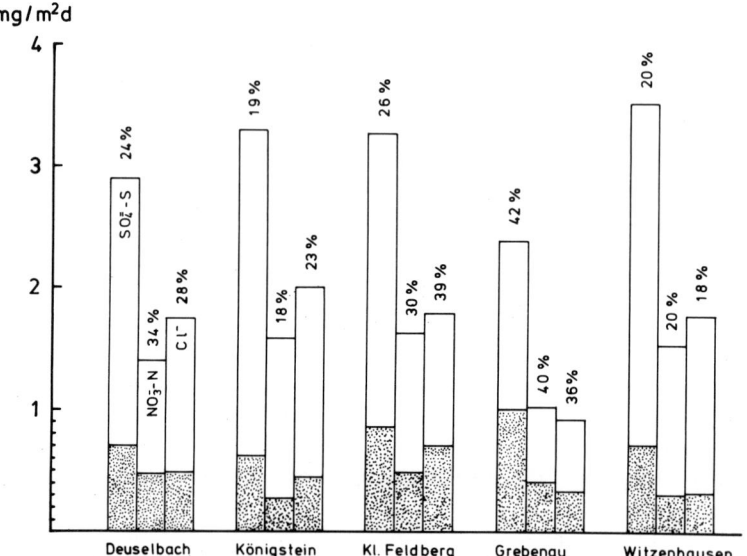

fig. 3: **Mean total deposition of anions. Percentage of dry deposition
shown as dark parts of the bars. SO_4^- and NO_3^- calculated as S
and N respectively.
(April 1983 - April 1985)**

Concerning the anions, the dry deposition with fractions of about 20 - 40 % is of minor importance at all stations. The total deposition of SO_4^- ranges from 2.4 to 3.5 mg/m²·d. The deposition rates of NO_3^- and Cl^- are in the same range with values of about 1 - 2 mg/m²·d. Concerning the heavy metals (fig. 4), obviously the fractions of wet and dry deposition are subject to wider variations with respect to space as well as element. Dry deposition occurs to be predominant for the elements Mn and Fe with fractions of about 40 - 80 % and about 70 - 90 % respectively. In opposite to that and similar to the anions, dry deposition is less important for Pb and Cd with fractions of 14 - 25 % (except of the station Grebenau with more than 40 %) and 11 - 39 % respectively. The trace metals described here, show total deposition rates that differ within 2 to 3 orders of magnitude. Highest deposition rates are reached by Fe with 290 - 540 µg/m²·d followed by Pb (17 - 38 µg/m²·d) and Mn (13 - 35 µg/m²·d). Cd reaches the lowest values of 0.7 -1.8 µg/m²·d.

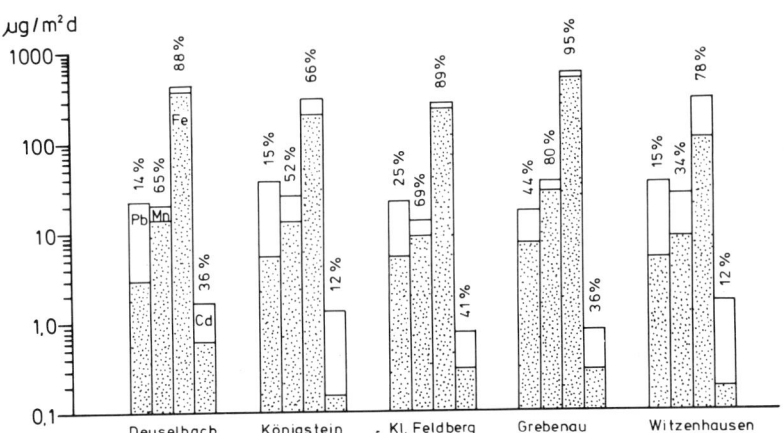

fig. 4: Mean total deposition of heavy metals. Percentage of dry deposition shown as dark parts of the bars. SO_4^- and NO_3^- calculated as S and N respectively.
(April 1983 - April 1985)

3.2 Episodic character of the wet deposition

As mentioned above, the wet deposition is dominated by the individual amount of rainfall. This nescessarily leads to the observed episodic nature. During these episodes, large fractions of the annual deposition of trace elements can occur within short periods of time. This phenomenon is of special interest when discussing possible damages caused by high deposition rates. After investigations through several years in Southern Scotland, Fowler and Cape (1984) stated that up to 40 % of the annual wet deposition of protones are related to less than 4 % of the annual rainfall-events. The data gained by the investigations presented here also document this episodic character. As an example, fig. 5 shows the number of rainfall-events at the station Königstein versus the

corresponding SO_4^- deposition arranged in order of decreasing deposition rates in a cumulative manner.
From this one can realize that in 1983 about 60 % of the rainfall-events were responsible for already 90% of the annual SO_4^--deposition. To put it in another way, 10 % of the rainfall-events are related to 40 % of the annual SO_4^--deposition.

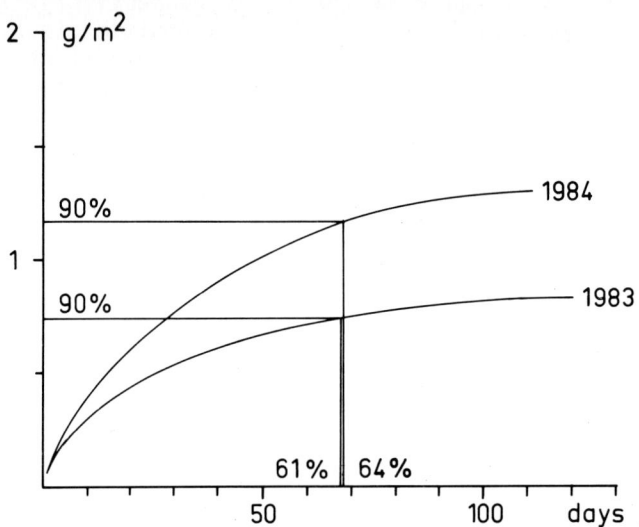

fig. 5: Cumulative wet deposition of SO_4^- caused by individual rainfall-events in dependence of the number of rainfall-events.

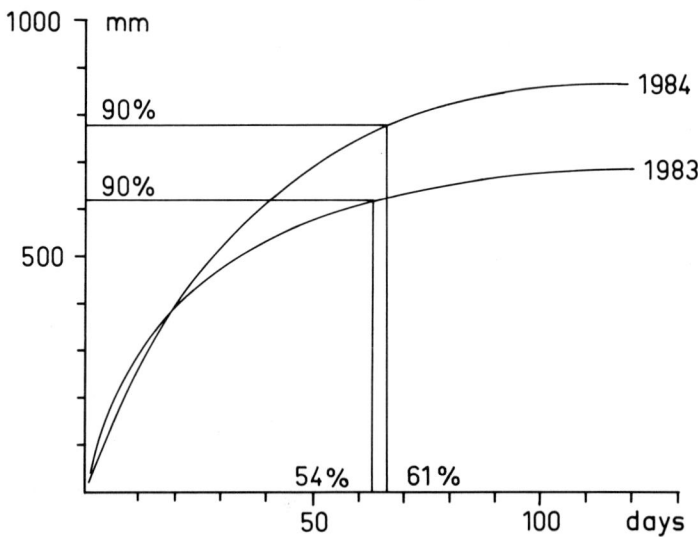

fig. 6: Cumulative wet deposition caused by individual rainfall-events in dependence of the number of rainfall-events.

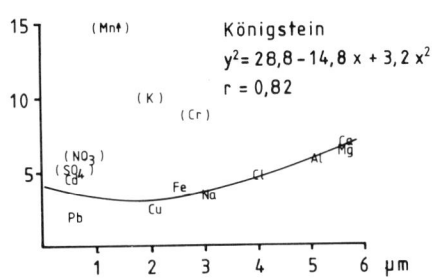
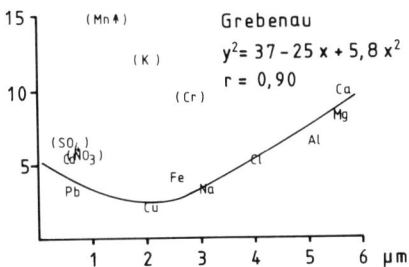
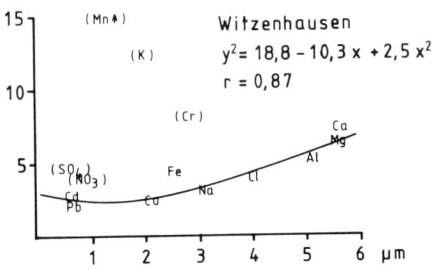

fig. 7: Concentration-ratios of precipitation samples (bulk-deposition) below and above canopies of three different forest stands in dependence of the mass median diameter of the corresponding aerosol particles

The measured concentrations of SO_4^-, NO_3^-, Cl^- and H^+ are presented in fig. 8. The standard deviation (shown as bars) of the mean concentrations given by the use of 9 collectors below the canopy indicate the inhomogenious conditions one has to take into account when sampling throughfall. The results from analysing the same precipitation sampled at a nearby clearing (adequate to sampling above the canopy), marked by crosses, show far lower concentrations.

In proportion, the concentrations below the canopy are enhanced by following factors:

SO_4^- : 7.8 Cl^+ : 13.3
NO_3^- : 21.0 H^+ : 10.1

In 1984, as expected, the same general behavior is observed. Anyway, we find a distinguished difference in the absolute amount of deposition rates with 0.83 g $SO_4^-/m^2 \cdot a$ in 1983 compared to
1.3 g $SO_4^-/m^2 \cdot a$ in 1984.
This is caused by a different frequency and amount of precipitation in both years documented in fig. 6. The rainfall rate at the discussed station Königstein was 688 mm/a in 1983 compared to 864 mm/a in 1984.

3.3 Deposition in the canopy

The canopy of trees to a great extent influences the potential amount of total deposition in forest stands. It represents a most complicated surface-area that gives rise to various deposition-processes, concerning gases as well as particles
Depositions, appearing on leafs and branches during periods of dry weather will be washed down with the following precipitation-event. The efficiency of this process depends on the amount and intensity of the rainfall.
Physically the deposition of particles depends on their size or more exactly on their inertia.
As a first approach measurements of the bulk deposition in forest stands in different levels over a longer period of time provide mean data suitable to estimate the significance of the dry deposition.
The mean ratio of element concentrations from these samples below and above the canopy plotted versus the mass median diameter (MMD) of the corresponding aerosol particles shows an increase with increasing MMD. This result should be taken as a first approximation due to the fact, that some possible processes taking place on the surface of the leaves especially in the region of the stomata are neglected. Fig. 7 shows this correlation to be found at all three forest stations in the same way. Certain elements appear with remarkable high values. Considerable enrichments in throughfall caused by the leaching of leafs respectively needles have to be taken into account in case of Mn and K (Höfken and Gravenhorst, 1982; Ulrich, 1983). Possible deposition of gases with following oxidation might influence the results in case of SO_4^- and NO_3^-. No explanation can be given in case of the unusual high ratio of Cr. Neglecting the components just mentioned, the relation will follow a mathematical function of type:

$$Y^2 = A + B \cdot x + C \cdot x^2$$

These results were obtained by using bulk-deposition samples at sampling periods of 14 days. Later on, "wet-only" sampling excluding direct dry deposition was applied at the station Königstein and daily sampling was established, improving the time resolution. Sampling like this should provide a more exact and more detailed information on the impact of a canopy on the throughfall only.
In autumn 1985 the occasion was arising to analyse precipitation-events following long periods of dryness. As one example, the rainfall-event of October the fifth following a dry period of almost four weeks was collected and analysed.

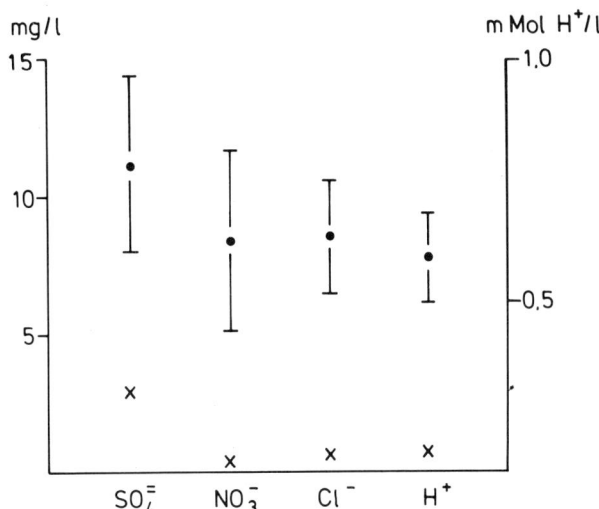

fig. 8. Concentrations in rain after a long period of dryness (4 weeks) below the canopy (•) and at a nearby clearing (x).
Sampling site: Königstein. Standard deviation occuring from sampling below the canopy (9 wet-only samplers) shown as bars.

4. DISCUSSION

The data obtained from the routine network show no significant differences between three forst stations and the remaining two stations. This is valid especially for the anions SO_4^-, NO_3^- and Cl^-. The mean pH values in precipitation varies within a small range of 4.3 - 4.4 representing conditions usually found in rural sites in the Federal Republic of Germany (Georgii et al., 1982). In recent investigations carried out in the northern part of the alps (Austria) as well as in the Kanton Zürich (Swiss), mean pH values of 4.3 to 4.6 and 4.4 respectively are reported (Puxbaum, 1985; Keller and Flückinger, 1985).
Varying fractions of dry deposition with respect to the total deposition have to be explained by different particle size distributions of the elements. Fe and Mn are mainly carried by large particles. Elements like Pb and Cd on the other hand to a large extent are bound on particles with 0.1 -1.0 μm in diameter. Particles within this size range are affected only little by sedimentation turbulent- and brownian diffussion. These particles therefore show the largest residence-times in the atmosphere.
Dry deposition rates derived from measurements by means of the Bergerhoff sampling system are not valid for natural conditions. Usually higher deposition velocities have to be expected on natural surfaces (Jonas, 1984).
The attempt applied by Höfken (1981) taking a canopy itself as natural surface is subject of further investigations
Thereby the dry deposition occuring within the canopy, also characterized as interception-deposition, is estimated essentially by calculating the

difference of concentrations from throughfall and unaffected precipitation.
One must be aware of problems and uncertainties due to the poor knowledge of interactions arising between leafs, dry deposition and rain. This includes processes like adsorption, chemical reactions and "leaching". By estimating the relevance of these effects, the relationship presented in fig. 7 should be verified.
Beside previous mentioned elements, Ca and Mg can be subject to the process of leaching. In case of the so called "leaf buffering", ions of Ca and Mg can be exchanged by protones, leading to an enhancement of these elements in throughfall not caused by interception-deposition (Ulrich and Matzner, 1983).
Nevertheless, neglecting these elements while calculating the best fitting graph in fig. 7, had almost no influence on the lapse as well as the quality of the adjustment.

Acknowledgements:

This investigation was sponsored by the "Umweltbundesamt" under contract no. 10 402 715. The author thanks the "Hess. forstl. Versuchsanstalt" and the "Hess. Landesanstalt für Umwelt" for technical assistance.

References:

Fowler, D., and Cape, J.N. (1984) 'On the episodic nature of wet deposited sulphate and acidity', Atm. Env. **18**, 1859-1866.

Gesellsch. für Strahl.- und Umweltforsch. (editor) (1985) Ergebnisse der Waldschadensforschung GSF-Bericht 8/85, München, 98 p..

Georgii, H.W., Perseke, C. and Rohbock, E. (1982) Feststellung der Deposition von sauren und langzeitwirksamen Luftverunreinigungen aus Ballungsgebieten, Abschlußbericht Forschungsprojekt 104 02 600 im Auftrag des Umweltbundesamtes, Eigenverl. des Universitätsinst. für Meteorologie und Geophysik, Frankfurt, 205 p..

Georgii, H.W., Rohbock, E. and Schmitt, G. (1984) Untersuchung des atmosphärischen Schadstoffeintrags in Waldgebieten der Bundesrepublik Deutschland, Zwischenbericht Forschungsprojekt 104 02 715 im Auftrag des Umweltbundesamtes, Eigenverl. des Universitätsinst. für Meteorologie und Geophysik, Frankfurt, 108 p..

Hess. Landesanst. für Umwelt (editor) (1984) Waldbelastungen durch Immissionen, 1. Zwischenbericht, Wiesbaden, 76 p..

Hess. Landesanst. für Umwelt (editor) (1985) Waldbelastungen durch Immissionen, 2. Zwischenbericht, Wiesbaden, 72 p..

Höfken, K.D. and Gravenhorst, G. (1981) <u>Untersuchungen über die Deposition atmosphärischen Aerosols an Buchen- und Fichtenwald</u>, Bericht im Auftrag des Umweltbundesamtes Forschungsprojekt 104 02 600, Eigenverl. des Universitätsinst. für Meteorologie und Geophysik, Frankfurt, 208p..

Keller, L. and Flückinger, W. (1985) 'Immissionsökologische Untersuchungen an Dauerbeobachtungsflächen im Wald des Kantons Zürich/Schweiz - Erste Ergebnisse der Beobachtungsperiode 1984/1985', <u>VDI-Berichte</u>, **560**, 253-289.

Jonas, R. (1984) <u>Ablagerungen und Bindung von Luftverunreinigungen an Vegetation und atmosphärischen Grenzflächen</u>, Diss. TH Aachen, Berichte der KfA-Jülich Nr. **1949**, 400 p..

Puxbaum, H. Pimminger, M. and Kovar, A. (1985) <u>Immissionsmessungen "Nasser Niederschlag" in Tirol</u>, Bericht 3 F/85 der Abteilung für Umweltanalytik am Inst. für Analy. Chemie der TH Wien, im Auftrag der Landesforstdirektion Tirol, Wien, Innsbruck, 69 p..

Rohbock, E. and Georgii, H.W. (1982) <u>Ein Depositionssammelgerät zur getrennten Erfassung der trockenen und feuchten Deposition atmoshärischer Schadstoffe</u>, Bericht im Auftrag des Umweltbundesamtes Forschungsprojekt 104 02 600, Eigenverl. des Universitätsinst. für Meteorologie und Geophysik, Frankfurt, 25 p..

Ulrich, B. and Matzner, E. (1983) <u>Abiotische Folgewirkungen der weiträumigen Ausbreitung von Luftverunreinigungen</u>, Forschungsbericht 104 02 615 im Auftrag des Umweltbundesamtes, Inst. für Bodenkunde und Waldernährung, Göttingen, 221 p..

VDI 2119 (1972) <u>Messung partikelförmiger Niederschläge, Bestimmung partikelförmiger Niederschläge mit dem Bergerhoff-Gerät</u>, (Standard-Verfahren), VdI-Handbuch Reinh. der Luft, Beuth Verl., Berlin.

PRECIPITATION INPUT OF INORGANIC CHEMICALS
IN THE OPEN FIELD AND IN FOREST STANDS
- RESULTS OF INVESTIGATIONS IN THE STATE OF HESSE -

H.M. BRECHTEL, Á. BALÁZS, F. LEHNARDT
Hessian Forest Research Station
Institute of Forest Hydrology
P.O. Box 1308
D-3510 Hann. Münden 1

ABSTRACT. The present situation is reported for open field precipitation input of 4 anions and 13 cations sampled during a state-wide pilot study covering three time periods totaling four months. Furthermore, results of comparisons between open area and adjoining spruce stands are presented. The annual average concentration and deposition rate for 1982 and 1983 are shown for 3 different locations in the Reinhardswald Research Area as well as for the time period October 1983 - September 1984 at several locations in the Research Areas Königstein, Grebenau and Witzenhausen.

1. INTRODUCTION

It is commonly accepted that a forest cover is normally the best natural protection for the chemical and biological quality of streams and ground water /19,20,21/. However, in light of the recent concern about the increasing environmental impacts of acidic immission input ("acid rain"), a significant change in this matter is suspected /22,27/. An example of increased acidic input is found in northern Hesse (Reinhardswald) where the 1975 open field sulfate input was three times greater and nitrate input twice as high as found 9 years earlier /31/. Thus, long before the overall immission impact on the forest ecosystem was generally recognized, it was already apparent that acidic immission and air pollution was no longer only a local problem of urban areas but also a factor which seriously endangers forested mountain watersheds remote from emission sources.
    In cooperation with a large number of research institutes and government agencies, the Institute of Forest Hydrology is working in several research projects concerning aspects of the immission impact on forest and forested watersheds /6/. Two of these studies began in 1972 /3,22/. This paper will report on 3 projects which started more recently to contribute to the clarification of the interrelationships of precipitation input of dissolved inorganic substances and forest stress. The concentration of elements or substances accumulated in open area or forest stand bulk precipitation samples at specific time

periods serves as a parameter for precipitation input.

First, some results of a state-wide pilot study will demonstrate the magnitude of concentration and monthly deposition of different chemicals found in the open area bulk precipitation collected in horizontal gauges throughout the State of Hesse. /2,5,7/.

Second, results from the Forest Hydrological Research Area of Reinhardswald and from 3 research areas of the Hessian Research Program "Forest Stress by Immission - WdI" /14/ are used to clarify the extent to which the trapping effect (filtering) can increase dissolved inorganic chemical concentrations and depositions reaching the forest floor in spruce stands.

2. MATERIALS AND METHODS

The state-wide pilot study /2,5,7/ was carried out in cooperation with the Federal Health Office in Berlin which analyzed the precipitation samples for Cl, F, NO_3, SO_4, Al, Pb, B, Cd, Ca, Cr, Fe, Cu, Mn, Na, Ni, Sr, Zn, (NO_2, PO_4, Be, Co and Mg). The substances in parentheses normally had concentrations below the lower detection limit of the used analytic methods, ionchromatography (IC) for the anions /18/ and plasma emission spectrometry (IC-AES) for the cations /34/. The bulk precipitation samples from approximately 150 open area gauges were collected during 3 random study periods : August 1981, February + March 1982 and February 1983. The second study period covered 2 months because of the lack of measurable precipitation in February 1982. The gauges were left in situ through March 1982. The pilot study encompassed the 18 measuring areas of the Hessian Forestry Precipitation and Snow Service Program /5/ and also the areas Hann. Münden City (excluded February 1983) and Hann. Münden vicinity (see Fig. 2 - 5). The precipitation gauging stations are generally located on northern and southern expositions at 100 m elevation intervals.

The measuring areas are:
    (1) Hinterer Odenwald, (2) Vorderer Odenwald,
    (3) FG Frankfurt, (4) Südlicher Taunus, (5) Hochtaunus,
    (6) FG Klingbachtal/Spessart, (7) Spessart,
    (8) Westlicher Vogelsberg, (9) Östlicher Vogelsberg,
    (10) Kuppige Rhön, (11) Hohe Rhön, (12) FG Krofdorf,
    (13) Nördliches Hessisches Schiefergebirge,
    (14) Kellerwald, (15) Knüllgebirge, (16) Meißner,
    (17) Habichtswald, (18) Reinhardswald,
    (22) Hann. Münden City, (23) Hann. Münden Vicinity.
FG = Forest Hydrological Research Area.

The second project covers the time period February 1982 till March 1984 in the Reinhardswald Research Area in connection with weekly stream water sampling in the Elsterbach-Experimental Watershed /19,20, 21,22/. Bulk precipitation samples were collected over 2 week intervals at open area, European beech and Norway spruce sites characterized by varying conditions: easterly exposure at 160 m a.s.l. (meters above sea level), valley floor at 220 m a.s.l. and southwesterly exposure at 365

- 430 m a.s.l. The primary analysis of Cl, NO3, SO4, PO4 and NH4 was done in the laboratory of the Hesse Forestry Research Station. Chloride analysis was done by titration and the remaining elements by spectral photometry. During the second year, in cooperation with the Hessian Agricultural Experimental Station in Kassel, Pb, Cd, Cu and Zn were additionally analysed by atomic adsorption spectrometry.

The third project is part of the Hessian Research Program "Forest Stress by Immission - WdI" /14/ which started in spring 1983 in 3 areas: Witzenhausen (east of Kassel), Grebenau (northeast of Giessen) and Königstein (northwest of Frankfurt). In 1985 a fourth and fifth area were included and in 1986 the sixth and final area will be established. Quantitative gaseous, solid and liquid immission input analyses are done for each of the Norway spruce and adjacent open area sites at the main monitoring stations. Additional research plots are located in the vicinity of the main stations to investigate the spatial variation of both open area and forest precipitation deposition (areas 1 - 3 only spruce, areas 4 - 6 spruce and beech). This paper reports on the areas Witzenhausen, Grebenau and Königstein where two week samples of bulk precipitation are taken from 8 open area and 11 Norway spruce stations. Analysis work was undertaken by the Meteorological Institute, University of Frankfurt, using the same methods as mentioned in connection with the state-wide pilot study.

The "Münden 100" bulk precipitation gauge /4/ was used to collect water samples. In forest stands, 10 to 20 random station duplications are needed to quantify the areal average of precipitation reaching the forest floor as throughfall and drip. As shown in Figure 1, the "Münden 100" gauge is composed of two connected polyethelene receptacles, one of which functions as a funnel and the other as a 2000 cm^3 accumulator. The bottom part of the funnel is filled with glass beads, both to reduce evaporation from the accumulator and to reduce pollution problems from solids. The glass beads are held in place by a perforated ceramic plate. The funnel is reinforced by a plexiglass ring to insure a collecting surface of exactly 100 cm^2. Till 1983 the gauge was supported by a pole as shown in the upper part of the picture. Then the gauge mounting was considerably improved as is seen in the lower part of the picture. The gauge is now placed in a 1.25 m long plastic stand pipe. The stand pipe is not only more stable and durable but also helps retard algal growth (limited light) and prevents evaporation by keeping the sample cool. In addition to the soft plastic net, a further improvement was necessary to avoid the problem of bird dung contamination. The best results to date have been found with a simple plastic ring 2.5 cm above and 7 cm from the collecting surface. No influence on the precipitation measurement has been observed. Comparative investigations indicate an excellent volume correlation between the "Münden 100" and the Hellmann Precipitation Gauge with a 200 cm^2 collection surface. This German Weather Service standard gauge cannot be used for water sample analysis because of its metal construction.

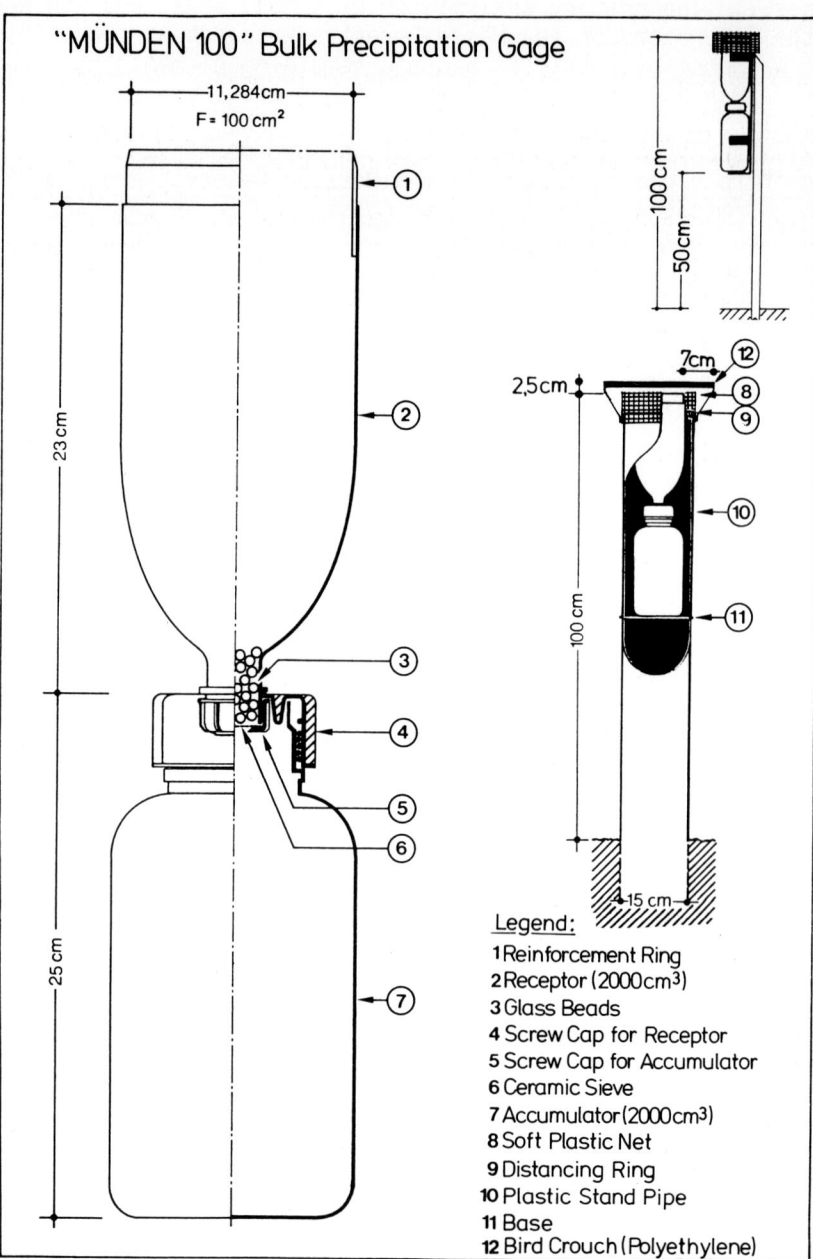

Figure 1: Construction plan of the "Münden 100"-Gauge /4/.

3. RESULTS

The investigations of the present immission impact of water soluble inorganic substances in precipitation concentrate on two objectives. The state-wide pilot study (project No. 1), based on discontinuous open area bulk precipitation sampling, was designed to furnish information concerning regional and locational differences in immission impact due to emission, orography, exposition and elevation. Furthermore, it was necessary to define the magnitude and range for a large number of elements found in the open field bulk precipitation as a guideline for future research.

The second objective was to quantify the influence of the trapping effect in various tree species, seasons and varying site conditions on the investigated elements.

3.1. Sampling of Open Area Precipitation throughout the State of Hesse

Table 1 lists the concentrations of dissolved substances (mg/l) in open field bulk precipitation collected in August 1981, February + March 1982 and February 1983. As indicated by the term bulk precipitation, the concentrations analyzed represent the input from precipitation as well as dust sedimentation /28/. Open area bulk precipitation possibly represents mostly wet deposition whereas the bulk precipitation collected under forest stands includes a considerable amount of substances trapped by the canopy (filtering effect) /33/. Additional enrichment of the net precipitation measured on the forest floor is caused by crown leaching /10/. The influences of the trapping effect and crown leaching have been investigated for Norway spruce at the Reinhardswald Forest Hydrological Research Area (project No. 2) as well as on "Forest Stress by Immission - WdI" (project No. 3) research plots.

The 4 anions and 13 cations having concentrations above the lower detection limit show a considerable range between the highest (Maximum) and lowest (Minimum) area mean values for the 19 study areas containing approximately 150 gauging stations. The variation coefficients for the 3 study periods are:

Cl	25 - 59 %	B	83 - 105 %	Mn	69 - 162 %
F	50 - 105 %	Cd	54 - 79 %	Na	103 - 118 %
NO_3	43 - 80 %	Ca	95 - 225 %	Ni	75 - 173 %
SO_4	35 - 51 %	Cr	77 - 239 %	Sr	61 - 235 %
Al	49 - 88 %	Fe	97 - 309 %	Zn	151 - 201 %
Pb	57 - 433 %	Cu	77 - 130 %	precipitation	26 - 35 %

In the case of dissolved aluminium, lead, manganese and zinc the highest mean value of a measuring area (Maximum) and, with lead and zinc, even the Hessian average of the open area bulk precipitation concentrations have been found in the magnitude of the tolerable highest concentration in drinking water /11/.

The lowest mean concentration (Minimum) from the various areas could be utilized to define "clean precipitation of Hesse" and deter-

Table 1: Concentrations (mg/l) of Dissolved Inorganic Chemical Substances in the Open Field Bulk Precipitation, Measured throughout the State of Hesse in 19 Different Measuring Areas (About 150 Gauging Stations), Maximum = Highest Mean Value of a Measuring Area, Minimum = Smallest Mean Value of a Measuring Area.

Substance, Symbol		August 1981 Ø	Max.	Min.	February 1982 Ø	+ March Max.	1982 Min.	February 1983 Ø	Max.	Min.
Chloride,	Cl	2.3	2.7	1.8	2.6	3.7	1.5	4.18	6.1	2.1
Fluoride,	F	<0.1	<0.1	<0.1	0.14	0.25	<0.1	<0.1	0.21	<0.1
Nitrate,	NO₃	4.0	7.3	1.9	6.2	9.4	3.8	2.74	5.1	1.0
Sulfate,	SO₄	5.5	7.3	3.6	9.9	15.5	5.7	6.5	10.5	3.4
Aluminium,	Al	0.11	0.15	0.08	0.14	0.27	0.05	0.14	0.25	0.04
Lead,	Pb	0.03	0.05	<0.02	0.05	0.08	0.03	0.02	0.25	<0.02
Boron,	B	0.009	0.018	<0.005	<0.005	0.014	<0.005	<0.005	0.008	<0.005
Cadmium,	Cd	0.0016	0.0027	<0.0015	<0.0015	0.0023	<0.0015	<0.0015	<0.0015	<0.0015
Calcium,	Ca	<1.0	1.3	<1.0	1.1	3.6	<1.0	<1.0	14.0*	55.7*
Chrome,	Cr	0.002	0.011	<0.002	<0.002	<0.002	<0.002	0.0025	0.007	<0.002
Iron,	Fe	0.063	0.133	0.018	0.134	0.221	0.018	0.076	0.35	0.005
Copper,	Cu	0.016	0.033	0.006	0.016	0.045	0.008	0.018	0.035	0.009
Manganese,	Mn	0.01	0.023	0.005	0.049	0.12	0.016	0.026	0.068	0.007
Sodium,	Na	2.1	3.9	<1.0	<1.0	2.4	<1.0	1.3	3.3	<1.0
Nickel,	Ni	0.008	0.023	<0.005	0.008	0.016	<0.005	<0.005	0.005	<0.005
Strontium,	Sr	0.02	0.03	<0.02	<0.02	<0.02	<0.02	0.026	0.157	<0.02
Zinc,	Zn	0.06	0.14	0.04	0.19	0.71	0.05	0.13	0.64	0.04
Precipitation mm		66	125	208	74	122	45	36	57	23

Tolerable highest concentrations in drinking water /11/ in mg/l: *apparently disturbed by liming

Cl: 200, F: 0.7 - 1.5, NO₃: 50, SO₄: 250, Al: 0.2, Pb: 0.05,
B: (1.0), Cd: 0.005, Cr: 0.05, Cu: (0.1), Fe: 1.5,
Mn: 0.05, Na: 175, Ni: 0.05, Zn: (1.0).

mine a kind of standard in order to define pollution. These are, in mg/l:

Cl	1,5	B	<0,005	Mn	0,005
F	<0,1	Cd	<0,0015	Na	<1,0
NO_3	1,0	Ca	<1,0	Ni	<0,005
SO_4	3,4	Cr	0,002	Sr	<0,02
Al	0,04	Fe	0,005	Zn	0,04
Pb	<0,02	Cu	0,006		

Figure 2 shows the mean area precipitation for the 3 study periods compared with the collective mean for all gauging stations throughout Hesse. The portions of the columnar diagrams in black represent area means greater than the Hesse average. A black dot next to the area number indicates that the sum value for all 3 of the measurement periods is greater than the Hesse average. Southern and central Hesse had the greatest amount of precipitation during the 4 months study because of the exceptional amount of rainfall in August 1981.

Figure 3 is a computer plot example of open area sulfate concentrations in precipitation from February + March 1982. The results are subdivided in the slope expositions south and north as well as in the elevation groups <300, 300 - 600 and >600 m a.s.l. One notable result not only for sulfate but also for several other substances is that within the total number of all measuring stations no definite anion or cation relation was noted for either exposition or elevation during the entire 4 months study period. In many cases there are contradictory results with one substance within the different study areas. In Figure 3 this is apparent for sulfate concentration when the neighboring areas No. 16 and 17 with No. 22 and 23 are compared. The reason for this is given by the different emission conditions. The Hann. Münden area (No. 22 and 23) is characterized by a large scale distal immission impact which is augmented by a local SO_2 emission source which dominantly affects the lower elevation zone regardless of exposition /35/. Figure 3 also illustrates that the highest sulfate concentrations are found in mountainous areas exposed to the southwesterly wind drift. A vague trend of increased concentrations from southwest to northeast is present. This trend parallels the predominating southwesterly wind drift and is approximately parallel to a topographic low which begins northeast of Frankfurt, crosses the Giessen and Kassel basins and terminates in northeastern Hesse. The increased concentrations at higher elevations found near the eastern border of Hesse (areas 11, 10, 16 and 17) are possibly a mixture of pollutional input from the southwest as well as from the east.

Figures 4 and 5 show, as an example for the other anions and cations, the concentrations and deposition rates of sulfate and lead as area means. As discussed with Figure 2, it is obvious that the immission impacts on Hessian mountain areas are strongly influenced by the emissions from the highly populated and industrialized areas in the Rhine-Main-Plain as well as in the Giessen and Kassel basins. This is not only true for chemical substances emitted as gas (HCl, HF, NO_x and SO_2) but also for heavy metals as, for instance, aluminum, copper

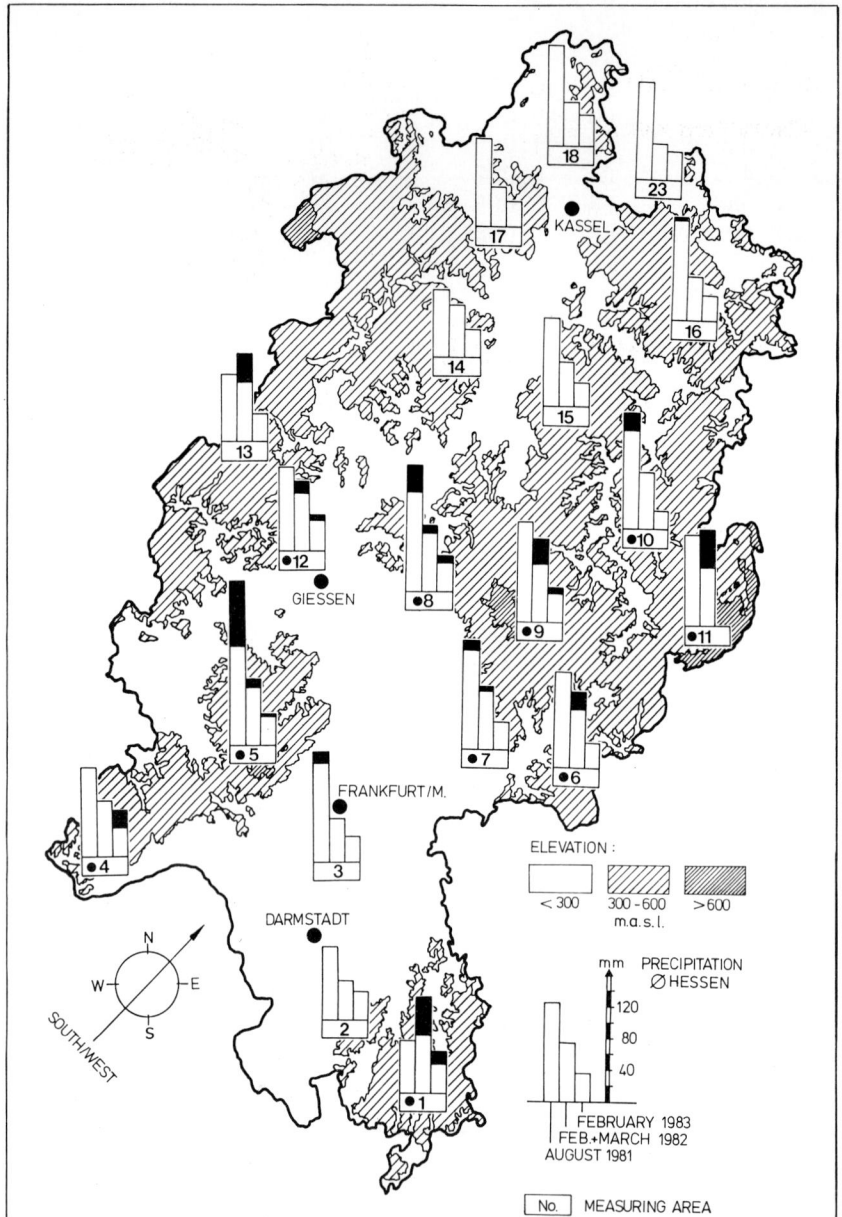

Figure 2: The conditions of **precipitation** during the study periods August 1981, February + March 1982 and February 1983, shown as area means (19 different areas, see chapter 2) in comparison with the total collective mean of all gauging stations (Ø Hesse), see Fig. 4 and 5.

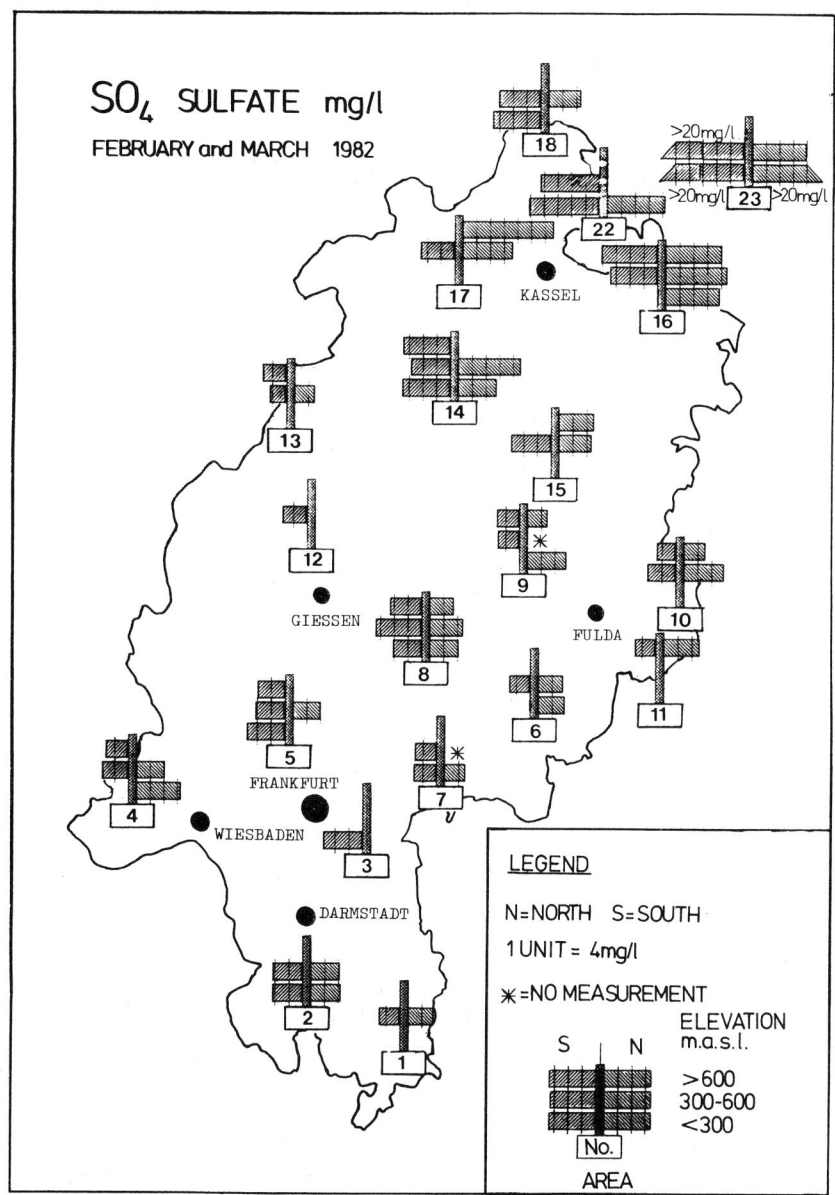

Figure 3: **Sulfate concentrations** of open area bulk precipitation samples, collected during the study period February + March 1982 on south and north slopes in the elevation zones <300, 300 - 600 and >600 m a.s.l.

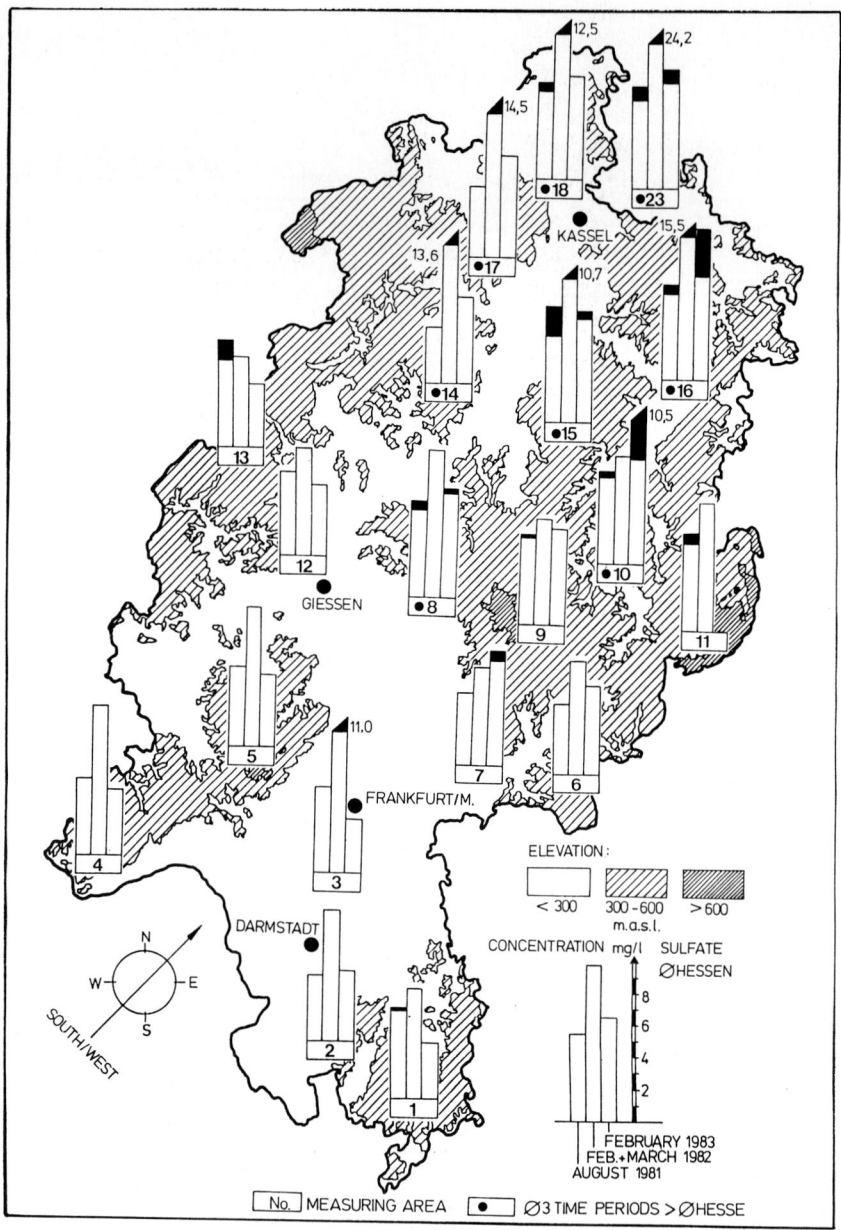

Figure 4a: Sulfate concentrations (mg/l) in open area bulk precipitation, shown as mean values for 18 measuring areas in Hesse and Hann. Münden vicinity during 3 different study periods.

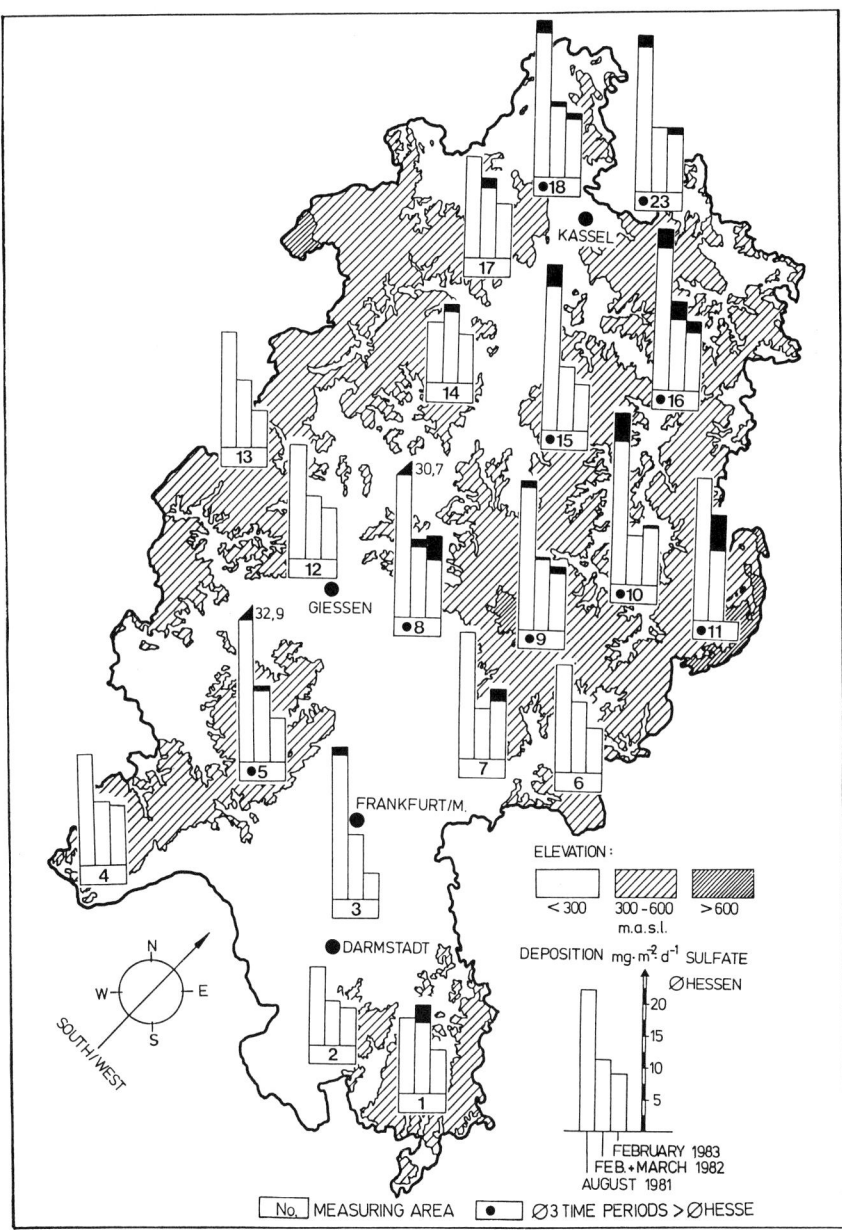

Figure 4b: Sulfate deposition rates ($mg \cdot m^{-2} \cdot d^{-1}$) in open area bulk precipitation. Wherever the area mean is greater than ∅ Hesse, the exceeding value is marked black in the columnar diagrams. Where the average of all 3 study periods exceeds the Hessian average, the area number is marked with a point.

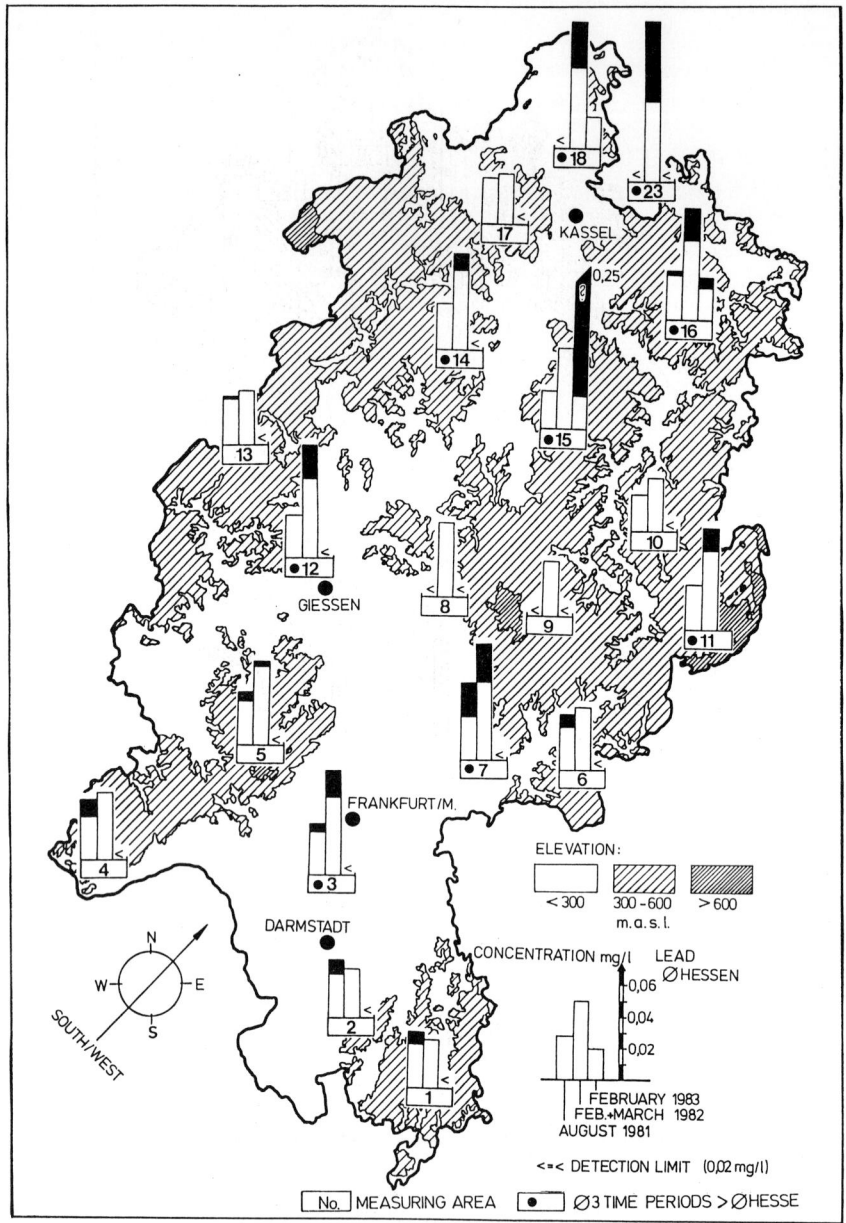

Figure 5a: Lead concentrations (mg/l) in open area bulk precipitation, shown as mean values for 18 measuring areas in Hesse and Hann. Münden vicinity during 3 different study periods.

PRECIPITATION INPUT OF INORGANIC CHEMICALS – RESULTS IN HESSE 59

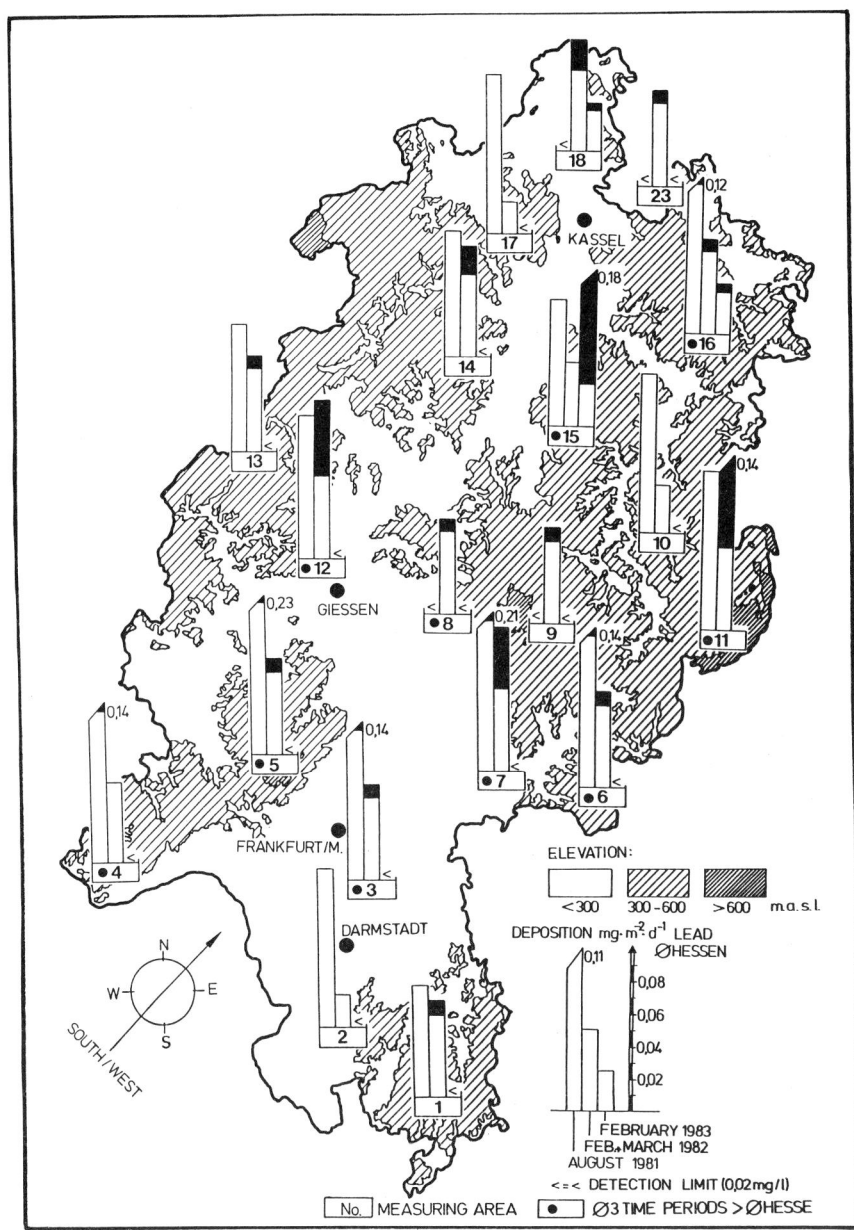

Figure 5b: Lead deposition rates $(mg \cdot m^{-2} \cdot d^{-1})$ in open area bulk precipitation. In comparison to sulfate (Fig. 4), where with the southwest wind a large scale distal immission is apparent, the distribution pattern of high lead values shows mainly proximal immission from local emission sources.

and zinc /2,5,7/. Thus, it can be concluded that even heavy metals are carried over far distances as aerosols. As far as lead (see Figure 5), cadmium, chrome, iron, nickel, strontium and zinc are concerned, the distribution pattern of high values shows a considerable influence of proximal immission from local emission sources /7/.

A rough estimation of the annual deposition rates of the 17 substances was made through the interpolation of the investigated 4 months time period values from the Hessian average. These interpolated values are, in kg/ha:

Cl	19	B	0,051	Mn	0,170
F	0,620*	Cd	0,010*	Na	12
NO3	33	Ca	21 (7)	Ni	0,051
SO4	50	Cr	0,015*	Sr	0;150*
Al	0,880	Fe	0,620	Zn	0,730
Pb	0,250	Cu	0,100		

* = the average concentrations for Hesse are below the lower detection limit (see Table 1).
( ) = probable value without February 1983 where some liming of sites resulted in excessively high concentrations.

Continuous sampling of open area bulk precipitation started throughout Hesse in spring 1983 as part of the above mentioned WdI-Research Program is used to verify the pilot study estimation.

3.2 Sampling of Net Precipitation in Norway Spruce Stands

In addition to the precipitation input of elements or substances in falling precipitation (rain, snow, hail, sleet) there is an additional input which depends on acceptors, like vegetation or other structures. This once again refers to the trapping effect, i.e. the incorporation of chemical substances in aqueous solution or suspensions like fog droplets (in this paper defined as a part of wet deposition) as well as the filtration of aerosols (a part of dry deposition) and the ab- and adsorption of gas on wet and dry surfaces (importance of intercepted water including dew and rime). A further factor is dust deposition and sedimentation /10/. Tables 2 and 3 illustrate the influence of the trapping effect and crown leaching of nutrients by Norway spruce on the concentration and deposition of dissolved inorganic substances in bulk precipitation.

Table 2 demonstrates the influence of different site conditions described in chapter 2 and annual variance of precipitation input of inorganic substances in 1982 and 1983 as concentrations and deposition rates in the Reinhardswald Forest Hydrologic Research Area. The average annual value (∅) for the 3 sites, as well as the highest (Max.) and lowest (Min.) site value for chloride, nitrate, sulfate, phosphate and ammonium concentrations and depositions are illustrated for both years. Only the 1983 annual average concentration and deposition is given for the heavy metals (Pb, Cd, Cu, Zn).

The upper part of the table compares the concentrations found

Table 2: Concentrations and Depositions of Chemical Substances Dissolved in Bulk Precipitation Collected in Reinhardswald.

I. CONCENTRATION (mg/l)

Substance, Symbol	Site	1982 ∅	1982 Max.	1982 Min.	1983 ∅	1983 Max.	1983 Min.
Chloride, Cl	Open	6.2	6.3	6.0	4.3	4.7	4.0
	Spruce	32.5	36.4	27.6	16.2	19.8	13.1
Nitrate, NO_3	Open	6.0	6.4	5.6	5.3	5.4	5.1
	Spruce	31.9	38.8	26.5	17.5	20.2	12.8
Sulfate, SO_4	Open	9.6	10.9	7.9	7.8	8.0	7.6
	Spruce	97.6	103.1	93.8	54.1	64.3	48.1
Phosphate, PO_4	Open	0.25	0.33	0.18	0.26	0.30	0.23
	Spruce	0.9	1.1	0.7	0.9	1.3	0.7
Ammonium, NH_4	Open	1.7	2.0	1.5	1.7	1.7	1.6
	Spruce	11.4	13.3	9.7	8.8	10.9	6.9
Precipitation mm	Open	623	654	582	774	804	742
	Spruce	317	445	248	396	490	322

II. DEPOSITION (kg/ha)

Substance, Symbol	Site	1982 ∅	1982 Max.	1982 Min.	1983 ∅	1983 Max.	1983 Min.
Chloride, Cl	Open	38.4	40.0	36.1	33.0	37.7	30.2
	Spruce	105.6	162.1	71.2	63.5	77.4	49.4
Nitrate, NO_3	Open	37.4	41.9	32.6	40.9	41.8	39.8
	Spruce	105.4	172.6	68.4	69.7	95.6	48.3
Sulfate, SO_4	Open	59.7	71.2	50.2	60.8	63.8	59.1
	Spruce	312.1	451.0	233.0	211.0	244.6	181.4
Phosphate, PO_4	Open	1.6	2.2	1.1	2.0	2.4	1.8
	Spruce	2.6	3.2	1.8	3.3	4.2	2.5
Ammonium, NH_4	Open	10.7	13.1	9.2	13.0	13.9	11.6
	Spruce	37.3	59.2	25.0	34.3	41.7	26.1

III. HEAVY METALS (1983)

	∅ Concentration (mg/l) Open	Spruce	∅ Deposition (kg/ha) Open	Spruce
Lead, Pb	0.02	0.047	0.137	0.186
Cadmium, Cd	0.0043	0.0068	0.033	0.027
Copper, Cu	0.01	0.08	0.086	0.301
Zinc, Zn	0.015	0.443	0.114	1.757

in the open field and spruce stands in relation to the annual sum of precipitation in 1982 and 1983. The average annual concentrations of the bulk precipitation input in the spruce stands are, with a much larger spatial and temporal variance, obviously much higher than in the open area /18/. The differences, expressed as multiples of the open area values, are:

Cl and NO_3 3 - 6, SO_4 6 - 12, PO_4 3 - 4, NH_4 4 - 7.

Only a certain element dependant part of the spruce stand enrichment is the result of a decreased net precipitation caused by an interception loss of 30 - 55 %. When expressed as multiples of the open area, the annual sum of deposition values (kg/ha) are:

Cl 2 - 4, NO_3 1 - 4, SO_4 3 - 6, PO_4 1.5 - 1.8, NH_4 4 - 7.

With reference to heavy metals analyzed in 1983 (bottom part of table), the multiple factors are for

Concentration: Pb 2.4, Cd 1.6, Cu 8.0, Zn 29.5
Deposition: Pb 1.4, Cd 0.8, Cu 3.5, Zn 15.4.

The annual mean concentrations for lead, cadmium and zinc in the spruce stands are within or above the highest tolerable concentrations for drinking water (compare with Table 1).
Table 3 illustrates the spatial variance of concentrations and depositions for the Research Areas Witzenhausen, Grebenau and Königstein. These stations form a NE-SW profile through Hesse. As described for Reinhardswald, the concentrations and deposition values in the spruce stands are considerably higher than in the open. The multiple values range for

concentration:				deposition:			
Cl	3 - 4	Ca	3 - 5	Cl	1.5 - 4	Ca	3 - 4
SO_4	4 - 6	Mn	20 - 70	SO_4	3 - 5	Mn	6 - 45
NO_3	3 - 4	Na	2.5 - 3	NO_3	1.5 - 3	Na	1.3 - 3.4
Pb	1.5 - 2	K	6 - 8	Pb	0.6 - (2.2)	K	3 - 9
Cd	1.5 - 3	Mg	2 - 5	Cd	1 - 3	Mg	2 - 4

( ) = unusually high values

It is interesting to note that the maximum lead and manganese concentrations in open area measurements are above the tolerable limits set for drinking water. Of greater concern is the fact that the annual average Pb and Mn concentrations for all 11 spruce sites are above these limits and not just specific location maximum values. This is indicative of a problem encountered when surface-near percolation water from spruce stands is influencing water resources which are for human consumption. In order to fully realize this problem, it must be furthermore considered that the measured soluble part of lead in the net

Table 3: Concentrations and Depositions of Chemical Substances Dissolved in Bulk Precipitation Collected in the Research Areas Witzenhausen, Grebenau and Königstein During the Time Period October 1983 - September 1984 (8 Open Area Stations and 11 Research Plots in Spruce Stands).

Substance, Symbol	Site	Concentration (mg/l) ∅	Max.	Min.	Deposition (kg/ha) ∅	Max.	Min.
Chloride, Cl	Open	1.5	3.0	0.9	13.5	28.6	6.2
	Spruce	4.6	6.2	3.8	28.2	38.7	17.5
Sulfate, SO_4	Open	5.8	6.6	5.0	50.2	65.7	37.1
	Spruce	28.2	35.9	25.1	172.8	221.7	124.0
Nitrate, NO_3	Open	3.3	4.0	2.8	28.6	34.2	20.0
	Spruce	10.6	15.5	7.4	65.1	86.5	34.3
Lead, Pb	Open	0.04	(0.08)	0.03	0.39	(0.75)	0.18
	Spruce	<u>0.07</u>	<u>(0.18)</u>	0.02	0.43	(1.05)	0.15
Cadmium, Cd	Open	0.0004	0.0008	0.0003	0.0028	0.0067	0.0023
	Spruce	0.0011	0.0018	0.0008	0.0065	0.011	0.0031
Calcium, Ca	Open	0.7	0.9	0.5	6.0	9.7	4.0
	Spruce	3.2	4.7	2.2	19.8	31.0	13.6
Manganese, Mn	Open	0.03	0.09	0.01	0.29	0.82	0.08
	Spruce	<u>0.88</u>	<u>1.8</u>	<u>0.26</u>	5.4	1.8	8.5
Sodium, Na	Open	0.7	1.0	0.4	6.0	8.7	3.1
	Spruce	1.8	2.7	1.5	11.2	16.7	5.7
Potassium, K	Open	0.4	0.6	0.3	3.7	5.8	2.3
	Spruce	3.4	5.0	2.7	20.7	37.6	14.5
Magnesium, Mg	Open	0.2	0.3	0.1	1.4	2.2	0.9
	Spruce	0.7	0.8	0.5	4.0	5.5	3.2
Precipitation(mm)	Open	871	1015	711	( ) = unusually high		
	Spruce	612	773	443			

<u>Tolerable highest concentrations in drinking water /11/ in mg/l:</u>
Lead: <u>0.05</u>, Manganese: <u>0.05</u>.

precipitation is only a part of the actual input because a considerable portion is also ab- and adsorbed in the forest canopy and is later deposited with litter /16, 24,25,26/. In connection with the increasing acidification of the forest floor by acid deposition, it must be expected that the long term accumulation of heavy metals in the litter layer and subsoil will redissolve and migrate with percolation water into surface near groundwater layers. Forest streams, characterized by a significant amount of subsurface flow and still considered to be clean may be confronted by increased pollution problems because of this factor.

4. CONCLUSIONS

The open area bulk precipitation in the State of Hesse contains concentrations of lead and, in some locations, other heavy metals which are higher than the tolerable limits set for drinking water. This is even more prevalent in forest stands and especially notable in spruce (Tables 1, 2 and 3). Also as indicated by the concentrations (mg/l) and depositions (kg/ha) of chloride, nitrate and sulfate, there is a serious impact of acidic input especially as far the stands of Norway spruce are concerned (Tables 2 and 3). Because of the trapping effects, the immission impact is much higher than in the open field. The results of investigations in the State of Hesse are verified by findings of numerous authors reporting from other States in the Federal Republic of Germany /1,12,13,15,23,29,30,32,33/. Acidic deposition in forest ecosystems has already attained levels at which deleterious effects on the soil matrix in some locations can be expected. This may alter, for instance, the fixation of nitrogen and cations in the forest soil so that nutrients and heavy metals may be leached by acidic seepage water into the ground water and into the streams of forested watersheds /9,17,22/. The high concentrations of heavy metals which have accumulated in the litter layer and subsoil can lead to serious problems /16,24,25, 26/. Future forest hydrology research should concentrate on the problem of water supply pollution caused by immission impact on the forest ecosystem, which, until now has been considered to be the best natural protection for water resources.

ACKNOWLEDGEMENT

The research reported in this paper has been accomplished in cooperation with the Federal Health Office in Berlin (Prof. Dr. Sonneborn), the Hessian Board for Environment (Dr. Boness, Dr. Hanewald) and Institute of Meteorology and Geophysics, University of Frankfurt (Prof. Dr. Georgii) and of course with help of many collegues in the Forest Research Station, especially Mr. Rapp, Miss Etta Starick and Miss Elke Starick. The authors express their sincere thanks.

REFERENCES

/1/ BLOCK, J.; BARTELS, U. (1985): 'Ergebnisse der Schadstoffdepositionsmessungen in Waldökosystemen in den Meßjahren 1982/83'. Forschung und Beratung, Reihe C, 39, Landwirtschaftsverlag GmbH, Münster-Hiltrup, 296 pages.

/2/ BRECHTEL, H.M. (1985): 'Deposition von Luftschadstoffen durch den Freilandniederschlag'. Allg. Forstz., 47, 1281 - 1283.

/3/ BRECHTEL, H.M.; BALAZS, A.; KILLE, K. (1982): 'Natural Correlation of Streamflow in the Forest Research Area of Krofdorf. - Results of a Paired Watershed Calibration'-. Proc. Symp. Hydr. Research Basins, Sonderheft Landeshydrologie, Bern, 291 - 300.

/4/ BRECHTEL, H.M.; HAMMES, W. (1984): 'Aufstellung und Betreuung des Niederschlagssammlers "Münden"'. Meßanleitung Nr. 3 der Hess. Forstl. Versuchsanstalt, Inst. f. Forsthydrologie, Hann. Münden, Eigenverlag, 13 pages.

/5/ BRECHTEL, H.M.; SONNEBORN, M. (1984): 'Gelöste anorganische Inhaltsstoffe in der Schneedecke unter Fichten- und Buchenbeständen und im Freiland der hessischen Mittelgebirge'. Deutscher Verband für Wasserwirtschaft und Kulturbau e.V., DVWK-Mitteilungen, 7, 527 - 543

/6/ BRECHTEL, H.M.; SONNEBORN, M.; LEHNARDT, F. (1985): 'Konzentration und Frachten gelöster anorganischer Inhaltsstoffe im Freilandniederschlag sowie im Bestandsniederschlag von Waldbeständen verschiedener Baumarten'. Reihe Tagungsberichte, 5, Nationalpark Bayerischer Wald, Grafenau, 153 - 168.

/7/ BRECHTEL, H.M.; SONNEBORN, M.; LEHNARDT, F. (1985): 'Räumliche und zeitliche Variation des Gehaltes anorganischer Inhaltsstoffe im Freilandniederschlag'. Verein Deutscher Ingenieure, Düsseldorf, VDI-Berichte, 560, 387-421.

/8/ BRECHTEL, H.M.; LEHNARDT, F.; SONNEBORN, M. (1986): 'Niederschlagsdeposition anorganischer Stoffe in Waldbeständen verschiedener Baumarten'. Schriftenreihe d. Dachverbandes Agrarforschung, München, 11, (in print).

/9/ BÜCKING, W. (1985): 'Einfluß von Bestand und Standort auf einige Bioelementgehalte im Sickerwasser'. Reihe Tagungsberichte, 5, Nationalpark Bayerischer Wald, Grafenau, 491 - 504.

/10/ DEUTSCHER VERBAND FÜR WASSERWIRTSCHAFT UND KULTURBAU (1984): 'Ermittlung der Stoffdeposition in Waldökosystemen'. Regeln zur Wasserwirtschaft, 122, 6 pages.

/11/ EUROPEAN COMUNITY COUNCIL (1980): 'Richtlinien des Rates vom 15. Juli 1980 über die Qualität von Wasser für den menschlichen Gebrauch (80/778/EWG)'. Amtsblatt der Europäischen Gemeinschaften, L 229/11 vom 30.8.1980, Brüssel, 11 - 15.

/12/ EVERS, F.-H. (1985): 'Ergebnisse niederschlagsanalytischer Untersuchungen in südwestdeutschen Nadelwaldbeständen'. Mitt. des Ver. f. Forstl. Standortkunde u. Forstpflanzenzüchtung, 31, 31 - 36.

/13/ FASSBENDER, H.W. (1984): 'Untersuchungen über die Deposition von chemischen Elementen in Waldbeständen in Bovenden bei Göttingen'. Forstarchiv, 1, 17 - 21.
/14/ GÄRTNER, E.J. (1983): 'Das hessische Untersuchungsprogramm Waldbelastungen durch Immissionen" - WdI'. Allg. Forstz., 38, 641 - 645.
/15/ HAUHS, M (1985).: 'Wasser- und Stoffinhalt im Einzugsgebiet der Langen Bramke (Harz). Dissertation, Forstl. Fak. d. Univ. Göttingen, Ber. d. Forschungszentrums Waldökosysteme/Waldsterben, 17, 206 pages.
/16/ HEINRICHS, H.; MAYER, R. (1980): 'The Role of Forest Vegetation in the Biogeochemical Cycle of Heavy metals'.Journal of Environmental Quality, Vol. 9, 1, 111 - 118.
/17/ HÜSER, R. (1979): 'Wald und Wasserqualität'. Schriftenreihe des DVWK, 41, 89 - 109.
/18/ KASISKE, D.; SONNEBORN, M. (1980): 'Analyse anionischer Bestandteile in natürlichen Wässern mittels Ionenchromatographie'. Labor Praxis, 4, 76 - 83.
/19/ LEHNARDT, F.; BRECHTEL, H.M.; BONESS, M. (1977): 'Nährstoffgehalte und Austräge von Bächen aus Einzugsgebieten verschiedener Landnutzung'. Verh. d. Ges. f. Ökologie, Göttingen 1976, Verlag W. Junk B.V., The Hague, 397 - 410.
/20/ LEHNARDT, F.; BRECHTEL, H.M.; BONESS, M. (1980): 'Wasserqualität von Bächen bewaldeter und landwirtschaftlicher Gebiete, - Untersuchungsergebnisse aus dem nordhessischen Buntsandsteingebiet -'. Forstw. Centralblatt, 2, 91 - 101.
/21/ LEHNARDT, F.; BRECHTEL, H.M.; BONESS, M. (1983): 'Chemische Beschaffenheit und Nährstofftransporte von Bachwässern aus kleinen Einzugsgebieten unterschiedlicher Landnutzung im Nordhessischen Buntsandsteingebiet'. Schriftenreihe des DVWK, 57, 177 - 298.
/22/ LEHNARDT, F.; BRECHTEL, H.M.; BONESS, M. (1984): 'Ein Beitrag zur Quantifizierung der Versauerung ausgewählter Bäche im Bereich des Nordhessischen Buntsandsteingebietes'. Materialien 1/84, Umweltbundesamt, Berlin, Erich Schmidt Verlag, 108 - 111.
/23/ MATZNER, E.; KHANNA, P.K.; MEIWES, E.; CASSENS-SASSE, E.; BREDEMEIER, M.; ULRICH, B. (1984): 'Ergebnisse der Flüssemessungen in Waldökosystemen'. Ber. d. Forschungszentrums Waldökosysteme/Waldsterben, 2, 29 - 50.
/24/ MAYER, R. (1983): 'Interaction of Forest Canopies with Atmospheric Constituents, - Aluminium and Heavy Metals -'. In: B. ULRICH and J. PANKRATH (eds), 'Accumulating Air Pollutants in Forest Ecosystems'. D. Reidel, Publ. Co., 47 pages.
/25/ MAYER, R.; HEINRICHS, H. (1980): 'Flüssebilanzen und aktuelle Änderungsraten der Schwermetall-Vorräte in Wald-Ökosystemen des Solling'. Zeitschr. Pflanzenernährung, Bodenkunde, 143, Weinheim, 32 - 246.
/26/ MAYER, R.; HEINRICHS, H.; SEEKAMP, G.; FASSBENDER, H.W. (1980): 'Die Bestimmung repräsentativer Mittelwerte von Schwermetall-Konzentrationen in den Niederschlägen und im Sickerwasser von Wald-Standorten des Solling'. Zeitschr. Pflanzenernährung, Bodenkunde, 143, 221 - 231.

/27/ MILDE, G. (1983): 'Grundwasserneubildung und Probleme der Gewässergüte'. Zeitschr. Dt. Geol. Ges., 134, 773 - 788.
/28/ MILLER, H.G., UNSWORTH, M.H.; FOWLER, D. (1980): 'Methods for Studying Acid Precipitation in Forst Ecosystems. - Definition and Concepts -'. Proceedings of a workshop held in Edinburgh 19. - 23. Sept. 1977, Institute of Terrestrial Ecology, 68 Hills Road, Cambridge, 36 pages.
/29/ RASTIN, N.; ULRICH, B. (1984): 'Depositionsmessungen in den Wäldern der Stadt Hamburg'. Ber. d. Forschungszentrums Waldökosysteme/Waldsterben, 2, 88 - 94.
/30/ REHFUESS, K.E. (1981): 'Über die Wirkung der sauren Niederschläge in Waldökosystemen'. Forstw. Centralblatt, 100, 363 - 381.
/31/ SCHEELE, G.; BRECHTEL, H.M.; SEIFERT, V. (1980): 'Immissionsbelastung durch den Niederschlag in Hann. Münden und im Reinhardswald'. Forstw. Centralblatt, 2, 91 - 101.
/32/ ULRICH, B. (1983): 'Interaction of Constituents. - SO_2, Alcali and Earth Alcali Cations and Chloride -'. In: B. ULRICH and J. PANKRATH (eds), 'Accumulating Air Pollutants in Forest Ecosystems', D. Reidel, Publ. Co., 47 pages.
/33/ ULRICH, B.; MAYER, R.; KHANNA, P.K. (1979): 'Deposition von Luftverunreinigungen und ihre Auswirkungen in Waldökosystemen im Solling'. Schriften Forstl. Fak. Univ. Göttingen u. Nieders. Forstl. Versuchsanstalt, 58, Frankfurt, 291 pages.
/34/ SONNEBORN, M.; BÄHN, U. (1982): 'Multielementbestimmung metallischer Bestandteile in Umweltproben mit Hilfe der Plasma-Atom-Emmissions-Spektrometrie (ICP-AES)'. In: B. WELZ (Hrsg.) 'Atomspektrometrische Spurenanalytik', Verlag Chemie, Weinheim.
/35/ ELROD, J.M. (1986): 'Investigation of Inorganic Pollutants in Bulk Precipitation in and around the City of Hann. Münden. Report of the Hessian Forestry Research Station, Institute of Forest Hydrology, Hann. Münden, 11 pages.

TRENDS OF HEAVY METAL POLLUTION BY WET DEPOSITION IN THE
FEDERAL REPUBLIC OF GERMANY DURING 1980 - 1984[*]

P. Valenta and V.D. Nguyen
Institute of Applied Physical Chemistry
Nuclear Research Center (KFA) Juelich
P.O.B. 1913, D-5170 Juelich
Federal Republic of Germany

ABSTRACT. Wet deposition of ecotoxic heavy metals Cd, Pb, Cu and Zn by rain and snow was followed in the FRG during the period 1980-84 to establish the long term trends. The common background heavy metal levels have not significantly changed over the studied period. The elevated depositions and concentrations in polluted regions, however, show since 1981 a decreasing trend. From the wet deposition data in unpolluted regions the background heavy metal pollution with precipitations in the FRG can be estimated.

1. INTRODUCTION

Ecotoxic heavy metals are emitted into the atmosphere in the FRG mainly from a variety of strong point sources and large diffuse sources. Thus, the most powerful Cd-emission sources are garbage burning, pit coal and brown coal burning, iron, steel and non ferrous metal production and glass industry whereas Cd-manufacturing contributes only by 2 % to the total Cd-emission. For Pb the largest emission source by far is automobile traffic, followed by iron and steel industry, coal burning and non ferrous metal production (Tab. 1) (Schladot et al. 1982; Schladot et al. 1986)

A major part of the emitted heavy metals constitute small dust particles (diameter of 0.1 - 5 µm) or gaseous organic species which remain for long time periods in the atmosphere as aerosols. They are transported by the wind over mesoscalic distances and are dissipated over the whole territory and bordering countries.

It has been established that for Cd, Cu, Zn and partly for Pb wet deposition with atmospheric precipitates by rain-out and wash-out is the predominant deposition mode, except in
[*] In memory of Prof. Nürnberg

the immediate vicinity of emission sources (Nguyen et al. 1979; Nürnberg et al. 1984). As the dissolved form of heavy metals is very favourable for their uptake by vegetation the wet deposition mode plays a very important role in terms of ecotoxicity.

TABLE 1
Estimated annual Cd- and Pb-emission in the FRG (Schladot et al. 1982; Schladot et al. 1986)

Emission source	t Cd/a	t Pb/a
Coal burning	11.8	450
Iron and steel production	7.7	2000
Non ferrous metal production	4.2	250
Traffic	0.5	3750
Garbage burning	11.1	5
Glass industry	5.0	–
Cd-manufacturing	0.9	–
Others	0.5	–
Total emission per year	42	6500

2. METHOD OF MEASUREMENT

Cumulative samples were taken on a weekly basis and monthly and yearly averages of the metal concentrations and depositions, resp., were calculated. A weighted mean of the metal concentration c and metal deposition d, resp., is defined as

$$c = \frac{c_t \, p_t}{p_t} \quad , \quad d = \frac{c_t \, p_t}{t}$$

where c_t, d_t and p_t are metal concentrations, metal depositions and the precipitation amounts determined in a cumulative rain water sample during the period t, i.e. a week.

The collection of relevant rain and snow samples was achieved by an automated wet-only sampler of own construction operating down to $-30^{\circ}C$ (Nguyen et al. 1981). The collected rain water or molten snow flows through a polyethylene funnel into a filtration device. Suspended matter is filtered off with a membrane filter of 0.45 μm pore size. The filtrate is collected in a precleaned polyethylene bottle.

Heavy metals Cd, Pb, Cu and Zn were determined simultaneously by differential pulse anodic stripping voltammetry (DPASV) at a hanging mercury drop electrode (Nguyen et al. 1979). Before the determination the sample was acidified to pH 2 and then subjected to UV-irradiation to decompose the inert metal complexes formed with the components of the dissolved organic matter. The determination limit lies at 50 ng/l, the relative standard deviation (RSD) is 5 % at concentrations 500 ng/l.

3. RESULTS AND DISCUSSION

3.1. Sampling network

A network of automated wet-only samplers distributed across the country (Fig.1) has been established in 1980 (Nürnberg et al. 1982). Here the results obtained in the semidecade

Figure 1. Sampling network

1980 - 84 are presented. Thus it was possible to follow the trends of the heavy metal pollution in typical selected regions, i.e. rural and sea-side regions without strong point or area emission sources (List/Sylt, Yerseke, Juelich, Deuselbach, Schauinsland, Hohenpeissenberg), polluted urban areas (Hamburg, Braunschweig, Frankfurt), the heavily polluted Ruhr region (Essen, Dortmund) and polluted regions with metallurgical industry (Goslar, Stolberg). Most of the samplers were located at stations of the German Meteorological Service or the Federal Environmental Protection Agency which also supplied meteorological data.

A substantial amount of the metals (80 % or more, excluding Pb) contained in aerosol particles is dissolved in the acid rain having a pH about 4. In the vicinity of an emission source, however, also the suspended matter contains appreciable amounts of heavy metals. It has been confirmed that the dissolved heavy metal concentrations in the rain are always significantly higher in the initial phase of rainfalls (Nguyen et al. 1979). Within 1 - 2 hours after the begin of a rainfall a substantially lower stationary metal concentration levels are attained in the rain. Thus the wet deposition of heavy metals depends not only on the metal concentration but also on the precipitation amount. This holds especially for the stations located on hills, e.g. Schauinsland and Hohenpeissenberg, with an elevated precipitation amount.

3.2. Wet deposition trends

With regard to the wet deposition of ecotoxic heavy metals the territory may be divided in a first approximation into various categories with common features (Tab.2). The areas of agriculture and forestry which correspond to about 85 % of the territory show a common anthropogenically influenced background range which does not change substantially with time and region. This can be seen from the rather stable values of the average daily wet deposition of Cd, Pb, Zn and Cu during 1980 - 84. This type of pollution is caused by a mesoscalic distribution of the heavy metal burden by atmospheric transport from the major emission zones in other parts of the country, e.g. the Ruhr region and urban agglomeration, or from the bordering industrialized countries. The contribution of the dry deposition to the total deposition will be not substantial in this region type. Thus the values given in Table 2 can also be used as a rough estimation of the average total heavy metal deposition in unpolluted regions of the FRG, i.e. in $\mu g/m^2 d$: 0.6 Cd, 30 Pb, 40 Zn and 7 Cu.

The category of urban agglomeration, e.g. Hamburg and Frankfurt area, shows for Cd and also for Pb and Zn an about twice higher deposition than in rural areas. Still higher deposition values were observed in the Ruhr region where the emission sources from industry, fossil fuel power plants and

TABLE II. Average daily heavy metal deposition with precipitates in µg/m²d in various region categories of the FRG during 1980 - 84

year	rural regions	Hamb.	Frankf. area	Essen	Dortm.	Stolbg.-Werth	Goslar
			Cd				
1980	0.4-1.0	1.0	1.2-1.4	2.5	2.5	1.7	6.5
1981	0.5-0.9	0.7	1.4-1.6	2.2	1.9	4.0	9.0
1982	0.3-0.7	-	-	1.9	1.0	2.4	5.3
1983	0.3-0.9	-	-	1.4	0.8	2.1	2.3
1984	0.3-0.9	1.0	-	1.5	0.9	2.1	1.6
			Pb				
1980	20-40	48	50-60	120	87	91	88
1981	20-40	38	50-80	150	100	200	150
1982	18-38	-	-	85	51	130	92
1983	15-40	-	-	50	34	85	110
1984	17-40	40	-	74	44	91	70
			Zn				
1980	25-70	81	80-90	220	240	73	520
1981	20-70	57	45-60	290	200	120	1530
1982	17-40	-	-	150	78	86	2020
1983	10-60	-	-	130	79	100	1105
1984	20-65	450	-	160	97	110	1520
			Cu				
1980	5-12	19	8-16	25	44	15	15
1981	3-15	13	13	25	26	27	30
1982	4-7	-	-	19	12	27	16
1983	3-7	-	-	17	9	25	14
1984	4-7	40	-	19	13	26	13

automobile traffic accumulate. The strikingly high value of the Zn-deposition and - to a lesser extent - of the Cu-deposition in Hamburg in 1984 is caused by the change of the sampling site from a residential part of Hamburg in 1980/81 to a new place in 1984 situated 6 km apart from the industrial zone in the prevailing wind direction.

The influence of the precipitation amount on the values of the Pb-deposition can be followed in Figs. 2, 3 and 4, where the average Pb-concentrations in the rain and snow, the average daily Pb-depositions and the average daily precipitation amounts, resp., are shown in polluted urban areas and in the Ruhr region during 1980 - 84. In general the pat

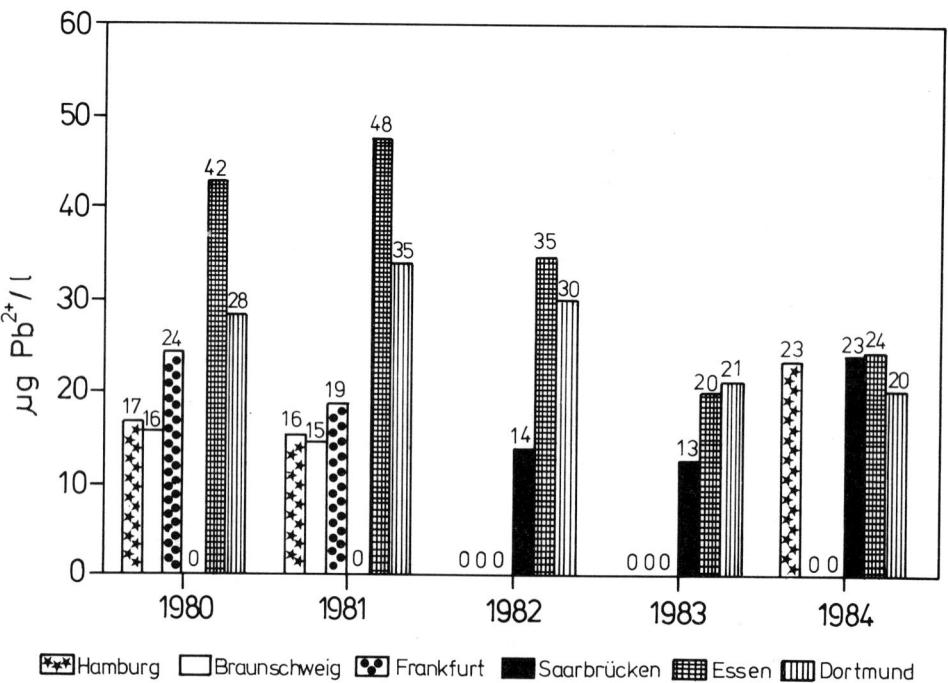

Figure 2: Average Pb-concentration (weighted yearly average) in polluted urban areas and in the Ruhr region. Sampling period 1980-84

terns in Fig. 2 and Fig. 3 are similar indicating that the elevated metal concentration values mainly determine the elevated metal deposition values. Nevertheless, there are some small differences, e.g. in Hamburg and Frankfurt in 1980, in Essen and Dortmund in 1983 and in Frankfurt, Saarbrücken, Essen und Dortmund in 1984, which are caused by the different average precipitation amounts at the sampling sites. The trends of the anthropogenically caused heavy metal pollution can be better followed from the metal concentration values (Fig.2) than from the corresponding wet deposition values (Fig.3) which are also influenced by the meteorological data. High deposition amounts of Cd, Pb and Zn are observed (Table 2) at the sites of lead smelters (Stolberg) and cadmium and zinc metallurgy (Goslar). In the vicinity of a strong point source, e.g. a large lead smelter in Stolberg, a substantially increased wet deposition takes place within a sector having typically a radius of several km around the emission source and decaying strongly with the distance from it. From our preliminary measurements in Stolberg results that in heavily polluted regions the dry deposition mode contributes substantially to the total deposition, especially for Pb.

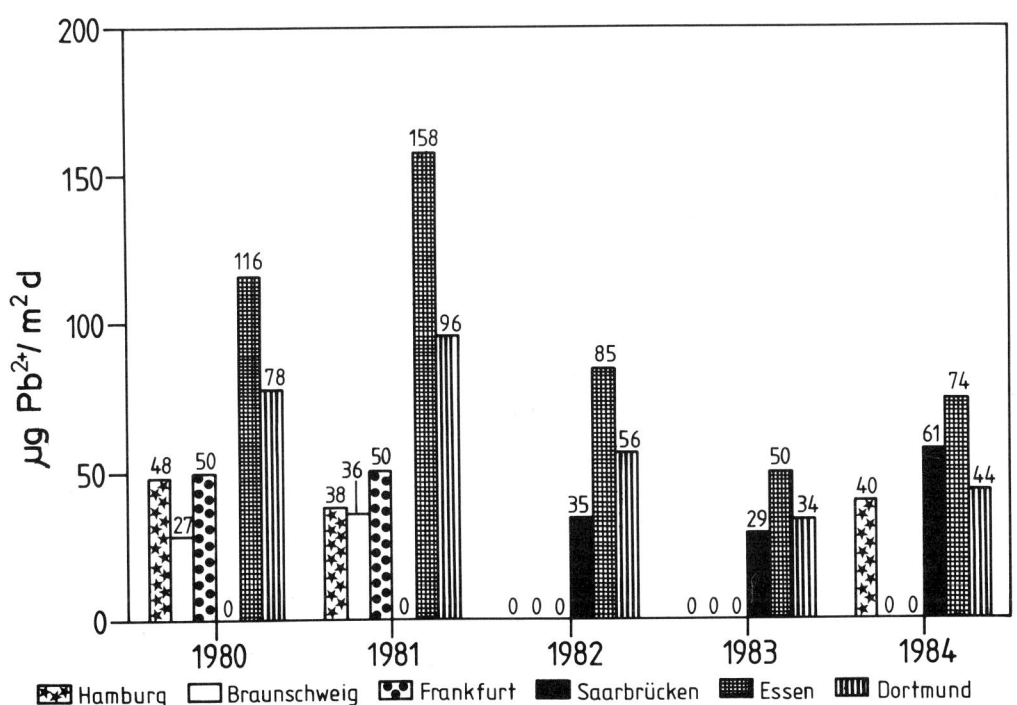

Figure 3. Average daily Pb-deposition with precipitates in polluted urban areas and in the Ruhr region. Sampling period 1980-84

Comparing these data with the rather stable deposition data in unpolluted regions a distint decreasing trend for Cd and Pb and partly for Zn since 1981 with a slight increase towards 1984 can be noticed in the Ruhr region and in Stolberg. This may be an indication of a decreasing pollution level in these heavily polluted regions in the last years.

The wet deposition data of heavy metals given here represent only the lower limit of the total heavy metal deposition as other deposition modes, i.e. dry deposition and interception, are not taken into account. Nevertheless the usefulness of the average heavy metal wet deposition data given here has been shown in a recent comparison between the estimated Cd-emission and the Cd wet deposition in the FRG (Schladot et al. 1986). If the Cd wet deposition is considered as the only Cd deposition mode an estimated yearly depo-

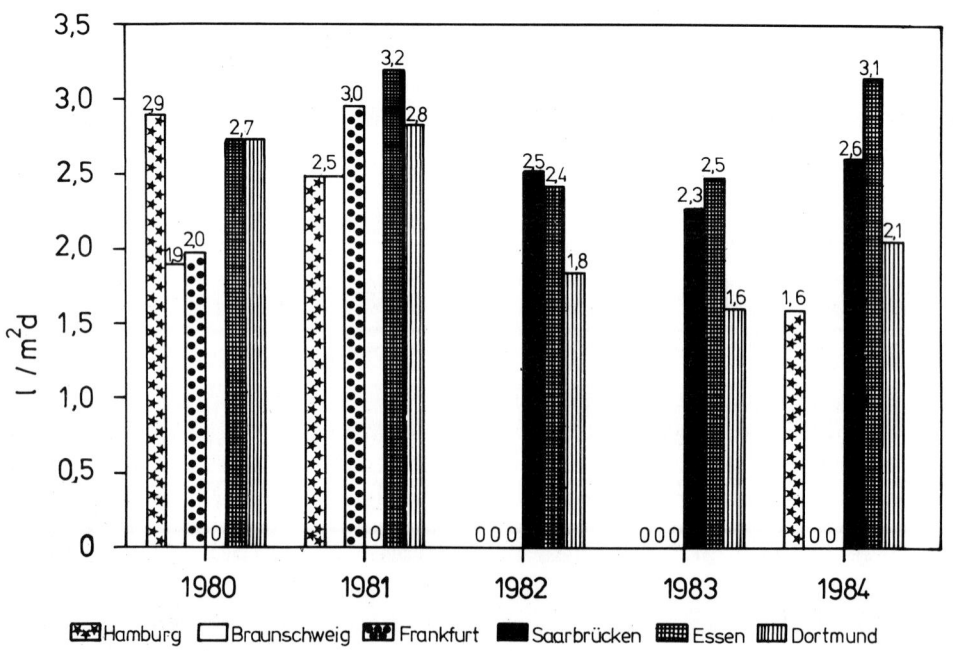

Figure 4: Average daily precipitation amount in polluted urban areas and in the Ruhr region. Sampling period 1980-84

sition of 58 t Cd over the whole territory of the FRG results, probably due to not precisely known Cd-emission factors and to omission of unknown Cd-emission sources. This value is, however, by about 30 % higher than the estimated yearly Cd-emission. This indicates - in spite of the approximative nature of the comparison - that wet deposition is by far the prevailing Cd-deposition mode in the FRG.

4. CONCLUSIONS

The heavy metal pollution by atmospheric precipitates in the FRG depends strongly on the region type. The common background pollution in unpolluted rural regions has not significantly changed over the measuring period 1980-84. On the contrary, the heavy pollution observed in some restricted regions (urban agglomerations, Ruhr region and regions with metallurgical industry) varies with time and has in general a decreasing tendency. As an estimate for the background heavy metal pollution with the precipitates in the FRG the following values result (in $\mu g/m^2 d$): o.6 for Cd, 30 for Pb, 40 for Zn and 7 for Cu.

ACKNOWLEDGEMENTS

The supply of meteorological data by the German Meteorological Service in Essen and the technical support of the meteorological stations of the German Meteorological Service and of the German Environmental Protection Agency is gratefully acknowledged.

REFERENCES

Nguyen, V.D., Valenta, P., Nürnberg, H.W. (1979).
 'Voltammetry in the analysis of atmospheric pollutants. The determination of toxic trace metals in rain water and snow by differential pulse stripping voltammetry'.
 Sci. Tot. Environm. 12, 151-167
Nguyen, V.D., Valenta, P. (1981) 'Gerät zur kontrollierten Probennahme von Regenwasser'
 USA Patent Nr. 4245499 (1981), Großbritanien Patent Nr. 2026178 (1983)
Nürnberg, H.W., Valenta, P., Nguyen, V.D. (1982) 'Wet deposition of toxic metals from the atmosphere in the Federal Republic of Germany', in: Deposition of Atmospheric Pollutants, Ed. H.-W. Georgii, J. Pankrath, D. Reidel, Dordrecht-Boston, p. 143-157
Nürnberg, H.W., Valenta, P., Nguyen, V.D., Gödde, M., Urano de Carvalho, E. (1984) 'Studies on the deposition of acid and of ecotoxic heavy metals with precipitates from

the atmosphere'.
Fresenius Z. Anal. Chem., **317**, 314-323
Schladot, J.D., Nürnberg, H.W. (1982) 'Atmosphärische Belastung durch toxische Metalle in der Bundesrepublik Deutschland - Emission und Deposition'
Jül 1776, KFA Jülich, 80 pp.
Schladot, J.D., Nürnberg, H.W. (1986) 'Cadmium Emission in the Federal Republic of Germany - Estimation of the atmospheric burden'
Sci. Tot. Environm., in press

BULK DEPOSITION INTO THE CATCHMENT "GROSSE OHE".
RESULTS OF NEIGHBOURING SITES IN THE OPEN AND UNDER SPRUCE AT
DIFFERENT ALTITUDES

G. Gietl*, A.M. Rall**

* Bavarian Forest Exp. and
Res. Station
- Forest Hydrology -
Schellingstr. 12
D-8000 München 40
West Germany

** Chair of Bioclimatology and
Applied Meteorology of the
University
Amalienstr. 52
D-8000 München 40
West Germany

ABSTRACT. In the catchment 'Grosse Ohe' in the National Park 'Bayerischer Wald' (φ = 48°57' N; λ = 13°26'E) quantitative forest hydrological studies are carried out since 1978. Since 1981 the pH-values of the precipitation in the open are measured, since 1983 also the bulk deposition at four sites in the open and two plots under spruce in different altitudes.

Upto now the results show that the spatial distribution of the bulk deposition in the open, independent of the temporal variation of the element concentrations, are caused by the distribution of precipitation in the catchment. Passing the canopy the precipitation is enriched by adsorption of gaseous deposition, filtering and leaching of air pollutions and nutrients. Exceptional high enrichments can be expected if the spruces are in fog or in drifting clouds filtering droplets. This process happens in about 50% of all sampling periods in the higher altitudes of the catchment.

1. THE CATCHMENT 'GROSSE OHE'

1.1. Site

The forest hydrological research basin 'Grosse Ohe' is situated in the National Park 'Bayerischer Wald' in the East of Bavaria. It has a size of about 19.1 km² and extends from the runoff gauge 'Taferlruck' (769 m a.sl.) to the peak of the 'Grosser Rachel' (1452 m a.sl.). Fig. 1 shows the geographical location and the shape of the catchment. Morphometrical, climatological and geological characteristics are described at IHP (1983).

Grosse Ohe is due to its location amidst widespread forests of the natural zone 'Hinterer Bayerischer Wald' a catchment far from industry; glassworks are the only important industrial plants in the near of the basin.

1.2. General objectives

About 98% of the catchment area are covered with forest (70% conifers:

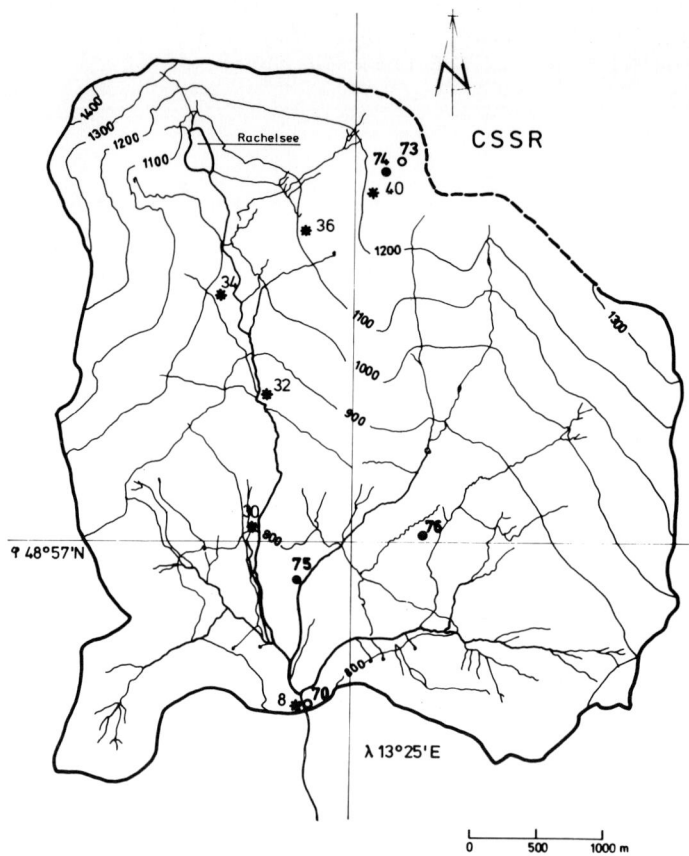

Figure 1. Location, shape and water system of the catchment Grosse Ohe. Inserted are the measuring sites for pH in the precipitation (*), the bulk deposition in the open (o) and under spruce (●).

spruce; 28% deciduous: beech and others) and will be brought back due to its location in the national park to a natural state. For that reason it has been declared as a biosphere reservation according to the MAB program and has been established in 1978 as a forest hydrological reference basin (Baumgartner, 1979).

The general aim of the hydrological investigations is to observe for a long period of time the water budget of a naturally wood covered catchment and to compare it to those of other basins exposed to human influences. For a shorter period of time special investigations are carried out to analyse the separate quantitative components of the water cycle. Examples hereto are publications concerning the determination of areal precipitation (Teichmann, 1984), the distribution of snow cover (Stang, 1986) and the application of a forest hydrological model (Kennel 1986).

Since water is also a transport medium for elements and particles, periodical investigations of the element cycle coupled to the water cycle are included in the program. Since 1981 measurements of the pH in precipitation as an indicator for pollution are carried out at various locations in the catchment, since 1983 measurements of the bulk deposition, too.

2. pH IN PRECIPITATION

2.1. Objective

The pH of precipitation may be considered in a first approximation as an indicator for the precipitation quality, especially for the content of acidifieres like sulfate and nitrate. Easily to determine the pH offers the possibility to measure it at several locations over a longer period of time as a parameter for the temporal and spatial variation of the bulk deposition.

2.2. Instrumentation

For the acquisition of a larger series of precipitation pH a teflon coated Hellmann rain gauge was installed at the base station Taferlruck (site # 8 in fig. 1) in the open emptied daily. For comparison purposes also bulk deposition samplers of the types Ulrich (Meiwes et al., 1983) and LÖLF (Block, 1983) were installed for the same time, emptied weekly. For exploring the spatial variation of the precipitation pH and the H^+ deposition five teflon coated rain gauges for daily sampling were installed in the open following a N-S profile (sites # 30, 32, 34, 36, 40 in fig. 1).

2.3. Results

A first evaluation of the pH measurement in the catchment Grosse Ohe (Gietl, 1985; Bosse, 1986) over the period from July 1981 to September 1984 shows the following results:

The weighted average of the pH at the station Taferlruck is about pH 4.2, the extremes pH 3.3 and pH 6.3 respectively. Depending on the amount of precipitation daily H^+ depositions range from 0.01 g/(ha*d) to 135 g/(ha*d). No pronounced annual variation of pH can be observed like at other stations (Kayser et al., 1974), though the weighted average of pH in wintertime (pH 4.3) is about 0.2 pH higher than in the summer (pH 4.1). The reason for this difference in the mean pH, which must be proved by means of additional measurements, may be caused by the different behaviour of sulfurdioxid diffusion in rain drops and snow respectively.

The measurements in different altitudes of the basin show no differences of the pH between the different stations on the average, although the values vary slightly on single days. The H^+ depositions, however, vary much more depending on the distribution of the local daily amount of precipitation and reached in the open a maximum value of 262 grams H+ per hectare and day.

The amount of precipitation measured by the daily emptied rain gauge agrees very well with that obtained from the weekly emptied samplers (types Ulrich and LÖLF). The samplers collect a 2% higher amount of precipitation than the rain gauges, which can be explained by higher wetting losses due to the daily emptying. The differences in the precipitation pH among the three types of collectors indicate that chemical reactions may occur more effectively in the samplers emptied once a week only. This leads to a rise of the lower pH in the weekly emptied samples, all in all to a higher mean pH value, and also to lower calculated H+ depositions compared to the values obtained from the teflon coated Hellmann rain gauge despite the difference in the precipitation sum.

3. BULK DEPOSITION

3.1. Objective

The aim of these studies is to quantify the amount of nutrients and pollutants deposited in connection with precipitation, because it is an important input into the element cycle of the catchment. Coincidently results in respect to the examination of the novel forest damages can be expected, which indicate to what extent the precipitation quality is a potential reason for the forest disease.

Since a long-termed precipitation quality study analogous to the quantitative hydrological program is not possible because of the great expenditure of laboratory, a first series of measurements is fixed on a duration of about three years. By means of a periodical repetition of this investigation in greater time intervals there is a chance of detecting trends in precipitation quality.

3.2. Instrumentation

For precipitation sampling in the open six collectors of the type LÖLF (Block, 1983) were installed at each of four sites in the open in different altitudes. Neighbouring to two of these stations two plots with 21 samplers were established in old spruce stands, having the possibility to estimate the deposition into the almost totally forested catchment by means of comparing the deposition in the open and under spruce. The locations of both pairs of sites are represented in fig. 1 with following numbers:

#	name	altitude	exposition	cover
70	Taferlruck	769 m a.sl.	horizontal	open
75	Grüben	790 m a.sl.	horizontal	spruce
73	Hochgfeichtet	1250 m a.sl.	11°, sw	open
74	Hochgfeichtet	1225 m a.sl.	11°, sw	spruce

In the meantime another plot is installed in a beech stand with additional instruments for sampling the stem flow.

All stations are instrumented following the standards of DVWK (1984); the sampling occurs weekly; pH, conductivity and water temperature are measured in the field, all of the other elements in the laboratory. Conservation, analysis methods and detection limits are described by Bosse (1986).

3.3. Results

The following results are based on the preliminary analysis of the first two years of measurement (25.7.1983-22.7.1985).

3.3.1 Mean annual deposition of elements. The stations for collecting the bulk deposition were equipped with the samplers in early summer 1983. After an operation on test of about eight weeks - meanwhile the samplers could age in a natural way - the routine measurement started

TABLE I.
Mean annual bulk deposition calculated from measurements between 25.7.1983 and 22.7.1985; comparison of different sites.

Element		Taferlruck 769m a.sl. 70 / OP	Grueben 790m a.sl. 75 / SP	Hochgfeichtet 1250m a.sl. 73 / OP	Hochgfeichtet 1225m a.sl. 74 / SP
H+	(kg/ha)	0.6	1.1	0.7	2.4
SO4	(kg/ha)	35.5	59.3	44.5	109.5
NO3-N	(kg/ha)	6.9	9.9	8.9	19.3
Cl	(kg/ha)	7.3	9.0	9.0	16.5
P total	(kg/ha)	0.45	0.93	1.08	0.99
NH4-N	(kg/ha)	6.5	5.0	9.6	9.9
Ca	(kg/ha)	3.9	10.7	5.6	21.0
K	(kg/ha)	2.4	13.9	2.5	16.4
Na	(kg/ha)	4.1	3.8	3.3	6.8
Al	(kg/ha)	0.3	0.7	0.4	1.2
Precipitation	(mm)	1105.5	830.1	1431.1	1315.5

OP: in the open   SP: under spruce

on 25.7.1983. Precipitation sampling in lower altitudes could be carried out all year long, the measurement at Hochgfeichtet had to be interrupted in the winter time because of the deep snow cover for several months. Thereby two complete years of measurement exist of Taferlruck (# 70) and Grüben (# 75), which do not correspond to the calendar year or the hydrological year. From these data mean annual values were calculated (tab. I). Because there exist only values of the summer months at the stations Hochgfeichtet (# 73, 74), the missing data of of the deposition in winter time (DW) are estimated from the ratio of the summerly deposition (DS) between lower and higher altitudes, separately for the sites in the open and under spruce, in a first approximation by the following equation for the individual elements (i):

DW (# 74,i) = (DS(# 74,i)/ DS (# 75,i)) * DW (# 75,i)
DW (# 73,i) = (DS(# 73,i)/ DS (# 70,i)) * DW (# 70,i)

The result is also presented in table I.

The table shows, that the precipitation in the open of the upper site is about 30% higher than at the base station, explainable by the difference of altitude. There is also a similar relation between the deposition rates at both sites. An exception occurs only at the element phosphorus which must be assumed to origin partly from local organic sources.

The comparison of both pairs "sites in the open : spruce stands" shows a reduction of precipitation in the stands relative to the open by interception losses. Concerning the elements an increase of deposition in the forest is seen on principle due to the following processes in the canopy: 1) adsorption of gaseous pollution (e.g. sulfur dioxide), 2) filtering of aerosols, 3) leaching of nutrients. These processes

Table II.
Bulk deposition sums (18.6.1984-8.10.1984) at different sites.

Element		Taferlruck 769m a.sl. 70 / OP	Grueben 790m a.sl. 75 / SP	Hochgfeichtet 1250m a.sl. 73 / OP	Hochgfeichtet 1225m a.sl. 74 / SP
F	(g/ha)	295.5 *	234.4 *	319.8 *	346.7 *
Mg	(g/ha)	321.3 *	643.8	370.6 *	898.2
Mn	(g/ha)	19.4	472.0	21.0	303.0
Fe	(g/ha)	74.3	165.0	87.7	291.9
Cu	(g/ha)	10.35 *	10.27	9.43 *	18.36
Pb	(g/ha)	32.0	21.2	31.9	22.3
Cd	(g/ha)	0.640	0.662	0.894	1.542
Hg	(g/ha)	0.686 *	2.431 *	0.896 *	1.142 *
Precipitation (mm)		503.5	379.7	622.3	582.1

OP: in the open   SP: under spruce
*: values unexact because of concentrations below the detection limit

differ to their extent fundamentally between the lower and higher altitudes and show a different enrichment for each element. Ammonium and phosphorus show also here an exemption being metabolized in the canopy and in the sampler by microbes (Ulrich et al., 1979; Müller et al., 1982). Adsorption and filtering cause an increase of the mean deposition by 50% to 100% in the horizontal stands in lower altitudes and by about 100% to 200% in the higher altitudes exposed to the wind. The enrichment of the nutrients varies according to the mobility of the elements and the acidity of the precipitation. Because of the higher acid deposition at Hochgfeichtet a greater leaching rate can be expected there. An exemption of that is shown by the element manganese only (tab.II

The chemical analysis of different elements, especially of the heavy metals, were stopped relatively early as their concentrations at all stations or at sites in the open were at or below the detection limit. For them no mean annual deposition rate according to table I could be evaluated. Therefore in table II deposition sums are listed for a shorter period of time only.

3.3.2. Special event. More definite hints concerning the causes of element enrichment during the canopy passage of precipitation is, based on the now existing instrumentation, available considering special events. During the fortnight from 27.8.1984 to 10.9.1984 the stations at Hochgfeichtet were for a long time exposed to fog and drifting clouds, causing an enhanced filtering effect. This can be documented by the amount of precipitation at the four sites (tab. III).

TABLE III.
Bulk deposition between 27.8.1984 and 10.9.1984 (Fog and drifting clouds at the Hochgfeichtet sites).

Element		Taferlruck 769m a.sl. 70 / OP	Grueben 790m a.sl. 75 / SP	Hochgfeichtet 1250m a.sl. 73 / OP	Hochgfeichtet 1225m a.sl. 74 / SP
H+	(g/ha)	50.8	122.2	50.4	290.6
SO4	(g/ha)	1851.5	5295.9	2716.2	10668.1
NO3-N	(g/ha)	144.9	452.8	231.4	1339.0
Cl	(g/ha)	201.2 *	366.8	251.5 *	733.6
P total	(g/ha)	8.1	19.6	10.1	44.3
NH4-N	(g/ha)	72.5	246.5	593.5	619.9
Ca	(g/ha)	80.5	725.2	100.6	1610.9
Mg	(g/ha)	40.3 *	100.1	50.3 *	138.9
K	(g/ha)	80.5	1304.3	100.6	1113.6
Na	(g/ha)	80.5	101.8	100.6	299.0
Mn	(g/ha)	0.4 *	122.1	0.5 *	57.5
Fe	(g/ha)	9.7	21.3	8.0	66.5
Al	(g/ha)	9.7	133.7	4.0	71.0
Cu	(g/ha)	4.03	2.02	1.01	4.13
Pb	(g/ha)	4.03	1.84	5.03	4.75
Cd	(g/ha)	0.081	0.123	0.101	0.495
Hg	(g/ha)	0.081	0.266	0.138	0.151
Precipitation (mm)		80.5	61.2	100.6	103.3

OP: in the open   SP: under spruce
*: values unexact because of concentrations below the detection limit

In the lower altitudes the spruce stand shows an interception loss of 19.3 mm or 24% respectively versus the neighbouring site in the open during that period. In higher altitudes the stand got 2.7 mm more precipitation than the site in the open in contrast to this. Assuming that this stand also had an interception loss of about 25%, it has a surplus of about 25 mm precipitation by filtering of cloud and fog droplets. As these droplets usually contain a higher concentration of substances than falling rain (Winkler, 1983) the bulk deposition under spruce at this station is significantly higher than in the annual average (tab. III, tab. I). The relationship of precipitation and deposition during that sampling period between the sites in the open (tab. III) corresponds to the average situation (tab. I) to a high degree, however, as the samples in the open collect only the falling precipitation. Ammonium and sulphate show compared to the ratio of precipitation too high values of deposition at Hochgfeichtet. Therefore an enhanced rain out of ammonium-sulphate aerosols at cloud level can be assumed.

Concerning the nutrients the high deposition rate under spruce caused by leaching mainly at the stand # 74 may be explained by the high deposition of acids. Contrary to the assumption that these substances are also filtered by the canopy there are the lower deposition rates (e.g. manganese) at the exposed stand in the higher altitudes, which cannot be explained aerodynamically but correspond well to the lower element concentrations in the needles.

Estimating the importance of filtering the fog and cloud droplets at higher altitudes the different sampling periods were evaluated according to the increase of filtering: Amongst a total of 39 samples six cases (= 15%) were recorded with a higher precipitation beneath the canopy compared to the site in the open; increases of concentrations caused by droplet filtering must certainly be assumed. At 13 other cases (= 33%) the interception loss at the upper stand was at least 50% lower than at the lower altitude; filtering is very likely.

3.4. Conclusions

The existing investigations on bulk deposition into the catchment Grosse Ohe have shown that the mean spatial distribution of element freights of precipitation in the open are caused at first by the areal distribution of precipitation. Passing the canopy the element concentration in precipitation rises by different processes. The enrichment factor depends on exposition and local weather conditions. An important part of bulk deposition in the spruce stand is caused b filtering cloud and fog droplets in. the upper part of the catchment.

Because the previous sample acquirement by totalizers offers only hints at the deposition processes, further studies in the open are estblished to sample and analyse fog and cloud water, to fractionate the bulk deposition into sedimentation and wet deposition and to observe the short-term variation of precipitation quality using a continuously pH recording tipping bucket. At least the project GASDEP (Enders; Teichmann, 1986) will present information on gaseous deposition when settled in the basin's area.

4. ACKNOWLEDGEMENTS

The studies described are part of the Project Grosse Ohe granted by the Bavarian State Ministries for Food, Agriculture and Forestry (project M 13) and for Education (project A 4). The authors would like to thank the collaborators of the Nationalpark Bayerischer Wald and those of the water board in Passau for the accurate sample acquirement and analysis as well as the technical staff of the Institute for instrumentation and service. We are also grateful to Prof. Dr. A. Baumgartner for guiding the project, Mrs. C. Mooslechner for drawing the figure and Mrs. B. Rasmussen for typing the paper.

5. REFERENCES

Baumgartner, A.: 1979. 'Zielsetzung des Symposiums "Wasserhaushalts-systeme naturnaher, kleiner Einzugsgebiete".' in: Nationalpark Bayerischer Wald, 1. Tagungsbericht, Grafenau. pp 12-16.

Block, J.: 1983. 'Pilotprojekt Saure Niederschläge. Beschreibung der Meßsysteme zur Ermittlung der Stoffdeposition in Waldökosysteme.' Landesanstalt für Ökologie, Landschaftsentwicklung und Forstplanung Nordrhein-Westfalen.

Bosse, M: 1986. 'Der Säure- und Stoffeintrag mit dem Niederschlag in das Einzugsgebiet Grosse Ohe unter Berücksichtigung von Jahreszeit und Witterungseinflüssen.' Thesis, Univ. of Munich, Dept. of physics. 131 pp.

DVWK, 1984. 'Ermittlung der Stoffdeposition in Waldökosysteme'. DVWK Regeln 122, Paul Parey Verlag Hamburg. 6 pp.

Enders, G. and Teichmann, U.: 1986. 'GASDEP-Gaseous Deposition Measurements of SO_2, NO_x and O_3 to a spruce stand: Conception, Instrumentation and first Results of an experimental project'. in: Proceedings of the Symposium Oberursel 1985, Georgii H.-W. (ed): D. Reidel Publ. Comp.

Gietl, G.: 1985. 'Der Stoffeintrag mit dem Niederschlag im Einzugsgebiet Grosse Ohe'. in: Nationalpark Bayerischer Wald, 5. Tagungsbericht, Grafenau. pp 431-444.

IHP, 1983. Hydrologische Untersuchungsbiete in der Bundesrepublik Deutschland. IHP/OHP Sekretariat Koblenz, Heft 4, ISSN 0724-7621.

Kayser, K., Jessel, U., Köhler, A., Rönicke, G.:1974. Die pH-Werte des Niederschlages in der Bundesrepublik Deutschland 1967-1972. DFG-Kommission zur Erforschung der Luftverunreinigungen, Mitteilungen IX, Verlag H. Boldt, Boppard.

Kennel, M.: 1986. Validierung, Anpassung und Modifizierung des forsthydrogischen Modells BROOK zur Simulation des Wasserhaushalts im Einzugsgebiet Grosse Ohe. Wasserhaushalt und Stoffbilanzen im naturnahen Einzugsgebiet Grosse Ohe, 3, Nationalpark Bayerischer Wald, Grafenau. (in print).

Meiwes, K.J. et al.: 1983. 'Methodik zur Erfassung der Deposition langzeitwirksamer Luftschadstoffe in Wäldern. Glatzel G. (ed), Hochschule für Bodenkultur, Wien pp 94-159.

Müller, K.P., Aheimer, G., Gravenhorst, G.: 1982. ' The influence of immediate freezing on the chemical composition of rain samples. in: Georgii, H.-W. Pankrath, J. (eds): Deposition of atmospheric pollutants. D. Reidel Publ. Comp. pp 125-132.

Stang, A.: 1986. Die Entwicklung der Schneedecke in den Wintern 1981/82 und 1982 und 1982/83 unter besonderer Berücksichtigung der Baumarten und Geländeausformung im Einzugsgebiet Grosse Ohe im Nationalpark Bayerischer Wald. Wasserhaushalt und Stoffbilanzen im naturnahen Einzugsgebiet Grosse Ohe, 2, Nationalpark Bayerischer Wald, Grafenau, 125 pp.

Teichmann, U.: 1984. Die Ermittlung des Gebietsniederschlages zur Lösung hydrologischer Bilanzen. Wasserhaushalt und Stoffbilanzen im naturnahen Einzugsgebiet Grosse Ohe, 1, Nationalpark Bayerischer Wald, Grafenau. 89 pp.

Ulrich, B., Mayer, R., Khanna, P.K.: 1979. Deposition von Luftverunreinigungen und ihre Auswirkungen in Waldökosystemen im Solling. Schriften aus der forstlichen Fakultät der Uni Göttingen und der Niedersächsischen Forstlichen Versuchsanstalt, 58, J.D. Sauerländers Verlag, Frankfurt.

INPUT OF ATMOSPHERIC POLLUTANTS IN A REMOTE HIGHLAND AREA

W. Kuttler
Faculty of Geosciences
Institute of Geography
Ruhr-University Bochum
P.O. Box 10 21 48
4630 Bochum 1, Federal Republic of Germany

ABSTRACT. In North Rhine-Westphalia, FRG, measurements of dry and wet depositions of the chemical constituents sulphate, chloride, nitrate, fluoride, calcium and ammonium were made in a remote highland (Sauerland, Rhenish Slate Mountains) and in an industrial area (Ruhr District). The measurements were carried out at 16 bulk-sampler stations and at two places with dry/wet samplers. Moreover the immission rates of sulphur and nitrogen compounds, either in the form of gas and particles, were determined. At the mountain station (845 m above sea-level) additional measurements concerning the wet deposition, dependent on the wind direction and the concentration of aerosol particles (for sulphate, nitrate and chloride) could be made. Estimations concerning wash-out and rain-out are given, as well as the percentages of the most important shares of acid-forming anions in the precipitation.

1. INTRODUCTION

For the registration of the immission and deposition structure of selected pollutants in the densely wooded, rural area 'Sauerland' (Rhenish Slate Mountains, North Rhine-Westphalia), dependent on space and time, measurements were made at 16 places between 1982 and 1984.

The area under investigation has an extent of 4,500 km^2, is predominantly covered by coniferous forests, between 300 m and 845 m above sea-level, and divided by numerous small rivers. In the valleys there are smallholdings, and here and there small industrial enterprises.

The data obtained in this area are compared to results obtained at the Bochum station in the industrial area. Fig. 1 shows the location of the measuring stations.

Bulk-samplers (SAM = surface active monitoring; after RUMPEL o.J.) were used as sampling stations, from which samples of the total monthly precipitation were taken; in addition these devices were equipped with filters, which allowed to determine the immission rate of sulphur and nitrogen compounds in the form of gas and particles (additional information in FUKUI 1966, KUTTLER 1985).

Additionally wet- and dry only samplers were installed on the highest mountain of the area under investigation (845 m above sea-level,

Kahler Asten; in the following called 'mountain station') and in the industrial area station (located in the Botanical Garden of Ruhr-University Bochum), to obtain samples of the diurnal total of precipitation.

Moreover, these two samplers allowed to measure the deposited dust and - with the sampler at the mountain station only - suspended dust particles.

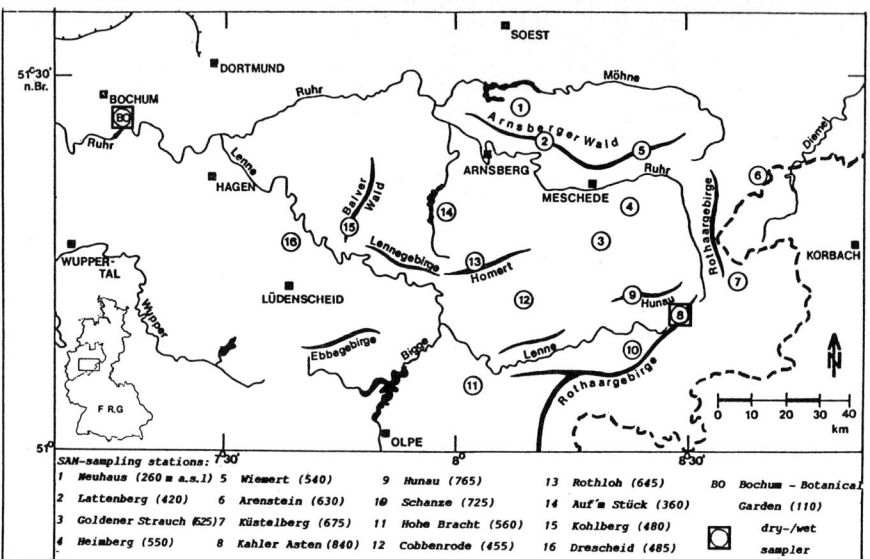

Fig. 1 Location of the measuring stations in the area unter investigation (Sauerland, North Rhine-Westphalia, FRG)

2. RESULTS OF THE MEASUREMENTS

2.1 Immission Rates

A survey about the spatial distribution of the immission rates of sulphur and nitrogen is given in fig. 2 and fig. 3.

The spatially limited character of the average sulphur-immission rates showed - for an ecologically unloaded area - a relatively high ratio of 1:4.6 between the place with the lowest deposition ($\bar{x}$ = 3.08 mg/m^2 · d at station 5) and that with the highest deposition ($\bar{x}$ = 14.31 mg/m^2 · d).

The spatial distribution of the average nitrogen immission rates (fig. 3) varies between 0.13 mg/m^2 · d (station 5) and 0.49 mg/m^2 · d (station 13). The ratio is 1:3.8. In general, a correlation coefficient of r = 0.94 (r^2 = 0.88) means a high correspondence between the average rates of the nitrogen and sulphur immission rates (exception: station 2, whose values were not taken into consideration, since there the sulphur rates were relatively high due to the emissions from the stack of a near small industrial source; cf. KUTTLER 1985).

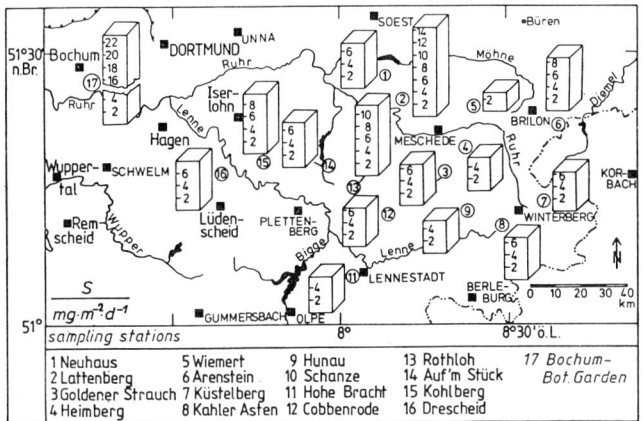

Fig. 2 Average values of sulphur immission rates in the remote highland area (Sauerland) and at Bochum station (industrial area) (12/1982 - 12/1983)

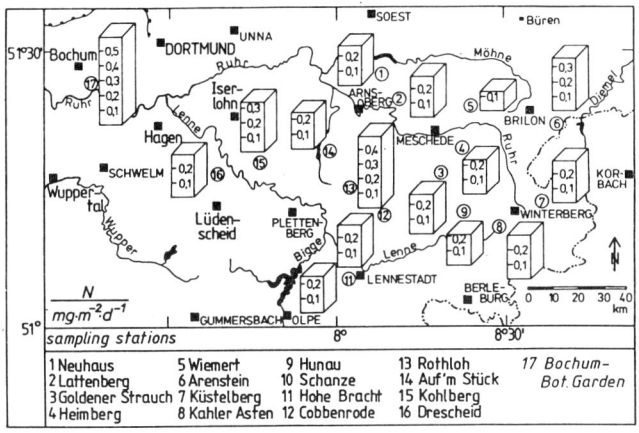

Fig. 3 Average values of nitrogen immission rates in the remote highland area (Sauerland) and at Bochum station (industrial area) (12/1982 - 12/1983)

Table 1 shows the seasonal differentiation of the immission rates.

In autumn and winter the sulphur immission at the stations in the remote highland as well as at the stations in the industrial area were higher than in spring and summer. During the winter months the regional average of the non polluted region was almost twice as high as the spring/summer average; this applies also to the station in the industrial area, where the rate is, however, three times higher.

The variation of the rates (lowest and highest annual average) for the individual season makes clear that even in the so-called non polluted regions relatively great differences in the immission structure - spatially and temporally - can be noticed; from the bioclimatic point of view this is an important aspect e.g. for the planning of health resorts. The highest values for nitrogen were recorded in summer, the lowest rates in autumn and winter. The seasonally adjusted values of the industrial area were twice as high as those of the non polluted region.

In contrast to the sulphur immissions, which vary especially in winter (1:5), the nitrogen immissions reached their peak level mainly in summer (1:6).

Tab 1. Sulphur and nitrogen immission rates in non polluted (remote highland area) and polluted (industrial) area (12/1982 - 12/1983) (in mg/m²· d)

season	sulphur remote highland area $\bar{x}$ amplitude[1]		sulphur industrial area $\bar{x}$	nitrogen remote highland area $\bar{x}$ amplitude[1]		nitrogen industrial area $\bar{x}$
winter	10,97	(4,7 - 23,7)	32,7	0,22	(0,17 - 0,34)	0,58
spring	5,41	(2,8 - 9,8)	17,6	0,29	(0,13 - 0,39)	0,53
summer	5,65	(2,8 - 9,3)	15,5	0,41	(0,13 - 0,78)	0,94
autumn	7,93	(2,3 - 16,0)	25,0	0,22	(0,11 - 0,53)	0,42
average	7,51	(3,1 - 14,3)	22,7	0,29	(0,13 - 0,49)	0,62

[1] highest and lowest average obtained by the station during the total measuring period

2.2 'bulk' - precipitation

The regional average of 'bulk'-concentrations as well as the corresponding 'bulk'-depositions (computed from 16 average rates obtained from the measuring stations) are shown in table 2 together with the comparative figures from the station in the industrial area.

Tab. 2. Weighted 'regional average values' of concentrations (in mg/l), of the pH-value and of the electric conductivity (in µ S/cm) as well as average depositions (in mg/m²· d) of sampling stations in the remote highland area (Sauerland) and industrial area (Bochum) (12/1982 - 12/1983)

	'bulk'-concentrations remote highland area $\bar{x}$	rel.dev. (%)	'bulk'-concentrations industrial area $\bar{x}$	'bulk'-depositions remote highland area $\bar{x}$	rel.dev. (%)	'bulk'-depositions industrial area $\bar{x}$
pH	4,22	2,6	4,24	-	-	-
electr. Cond.	43,00	16,1	58,00	-	-	-
SO_4^{2-} - S	1,64	17,7	2,56	4,81	13,3	5,80
NO_3^- - N	0,64	17,6	0,85	1,89	17,5	1,89
Cl^-	1,62	19,5	2,93	4,62	28,2	6,97
F^-	0,117	35,9	0,178	0,339	34,5	0,404

While the pH-average rates of the polluted and non polluted area hardly differ, the concentration rates for sulphate exceed the rates of the non polluted area by 56 %, those for nitrate by 33 %, those for chloride by 124 %, and those for fluoride by 52 %.

In the non polluted highland area the variations in relation to the regional average were less than 20 %; only for fluoride a higher rate of 36 % rel.dev. was found.

The values for the deposition rates show that compared to the industrial area rates in the non polluted area are either the same (in case of nitrate) or higher.

E.g. for sulphur (approx. 21 %), chloride (approx. 51 %) and fluoride (approx. 19 %). It must, however, be considered that the annual total precipitation in the clean air region is generally higher than in the industrial area.

In the non polluted area the deposition rates varies between 13 % and 28 % from the regional average value. Only the deviations of the fluoride input reaches 24 %.

2.3 'wet- and dry-only' samples

As far as 'wet-only'-concentrations are concerned it can be stated that the concentrations in rainwater measured by the mountain station differ considerably from those measured by the industrial area station (tab. 3).

Tab. 3. 50 %-percentages of 'wet-only' concentrations measured at a mountain station in the non polluted highland area (Kahler Asten, 845 m a.s.l.) and a station in the industrial area (Bochum, 140 m a.s.l.) (1/1983 - 2/1984)

	pH	el.Cond. (μS/cm)	SO_4^{2-}-S (mg/l)	NO_3^--N (mg/l)	NH_4^+-N (mg/l)	Cl^- (mg/l)	Ca^{2+} (mg/l)
industrial area	4,15	55,0	2,3	0,64	1,30	2,0	1,40
mountain station	4,17	30,0	1,2	0,44	0,81	0,8	0,42

To determine the frequency distribution of the concentration in the precipitation (for sulphate) at the mountain station and in the industrial area the sulphate concentrations (C) measured in the diurnal precipitation samples were compared to the thus computed average concentrations ($\overline{C}$). The figures obtained were subjected to a frequency distribution (fig. 4).

In this context cf. SMITH (1983), who defined an episode of short, but high load with the arbitrarily fixed 3 C-rate (= triple concentration of the average rate $\overline{C}$).

The results obtained here show that a 3 C-rate occurs in 1.5 % of all cases at the mountain station and in 2.5 % of all cases at the industrial area station.

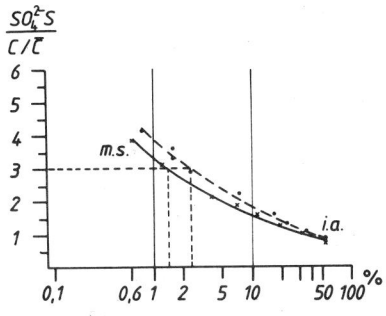

Fig. 4 Ratios between SO_4^{2-} - S concentration of diurnal precipitation samples (C) and the average rates ($\overline{C}$) computed thereof for the station in the industrial area and for the mountain station (1/1983 - 2/1984)

SMITH (1983) found that in Norway this rate is exceeded in 7 % of all cases.

The wet-only depositions are neither evenly distributed within the year nor within the month, but vary in some cases extremely in the first line due to the varying amount of precipitation, and secondly due to the varying concentrations of chemical constituents in the precipitation.

These temporally different records can result in a stress for the ecological systems, especially in the forests.

For sulfate, for example, therefore either the monthly percentage of deposition with regard to the total deposition during the measuring period and the daily maximum records were computed (fig. 5).

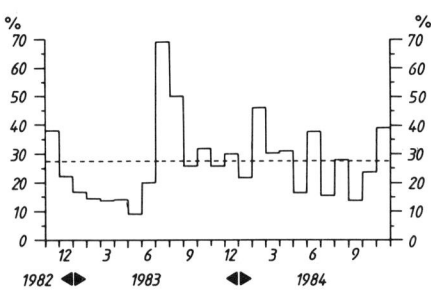

Fig. 5 Maximum diurnal share of SO_4^{2-}- S depositions of the monthly 'wet-only' total deposition during the measuring period at the mountain station in the non polluted area (11/1982 - 11/1984)

The average daily maximum sulphate-sulphur deposition - expressed in percentage of the monthly total - was 27.4 %.

The highest diurnal maximum was 70 % of the monthly deposition in July 1983. Extremely high deposition rates occur in connexion with short heavy showers.

Also from month to month you can record - referring to the deposition rate of the total measuring period - extremely varying figures due to the 'wet-only'-deposition (fig. 6).

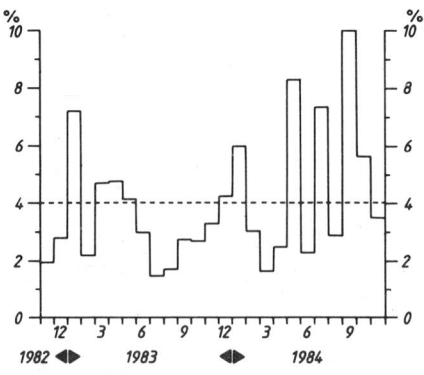

Fig. 6 Monthly percentage of SO_4^{2-}- S deposition of the 'wet-only' total deposition during the measuring period at the mountain station in the non polluted area (11/1982 - 11/1984)

So on one hand extremely high rates (10 % of the total deposition during the measuring period), and on the other hand extremely low rates (2 %) were obtained.

2.3.1 'Wet-only'-inputs depending on the wind direction. To determine the input of pollutants depending on rain-bringing wind directions the precipitations collected with the wet and dry only sampler on a diurnal basis at the non polluted mountain station were compared to the diurnal average of the eight main (surface-) wind directions. The results for the chemical constituents sulphate and chloride shall be explained (for further constituents cf. KUTTLER 1985).

From fig. 7 can be seen that the sulphate-sulphur concentrations dependent on the wind direction show no significant differences between the individual sectors; on the average the concentrations vary between 1.4 mg/l for precipitation from southwest and slightly higher values of 2 mg/l for precipitation from east and northeast.

A completely different distribution was recorded for the chloride concentration in the precipitation. The highest rates were obtained from rain-bringing winds coming from northwest (2.4 mg/l). Concentrations in the precipitation from the sector north-east to south showed the significantly low values of 0.5 to 0.6 mg/l.

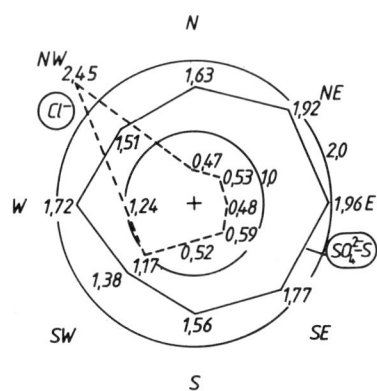

Fig. 7 Distribution of the weighted concentrations in rain dependent on the wind direction for SO_4^{2-}- S and Cl^- (in mg/l) at the mountain station in the non polluted area (11/1982 - 11/1984)

Fig. 8 shows the average chloride and sulphate-sulphur deposition rates for each sector. Sulphate reached the highest average in the east-sector. The reason is that only a few, but very heavy showers within the framework of some Vb-weather situations caused the high deposition rates.

In the case of chloride, however, the winds from northwest dominate with 31.4 mg/m² · d. The sectors north, east and south are obviously underrepresented: the input of particles from these directions showed rates of 3.1 to 7.9 mg/m² · d of chloride.

Compared to the total measuring period (11/1982 - 11/1984) - due to the main rain-bringing directions west and southwest - more than 65 % of the particles were recorded from these sectors.

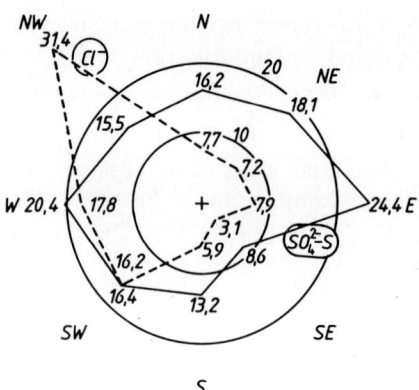

Fig. 8 Distribution of wet deposition dependent on the wind direction for SO_4^{2-}- S and Cl^- (in mg/m²·d) at the mountain station in the non polluted area (11/1982 - 11/1984)

2.3.2 <u>Aerosol Concentrations</u>. Aerosol concentrations of sulphate, nitrate and chloride particles were measured at the mountain station by means of a membrane and a teflon-filter (pore size 0.45 µm) respectively 0.2 µm) with an average suction rate of 1 m³/d for a period of 14 days. The individual data can be seen in table 4.

Tab. 4. Means of aerosolconcentrations at the mountain station in the remote highland area (Kahler Asten, 845 m a.s.l.) (in µg/m³) (11/1982 - 11/1984)

	$\bar{x}$	max.	min.	rel.dev. (%)	n
SO_4^{2-}-S	1,25	2,77	0,23	0,61	50
NO_3^-- N	0,35	0,75	0,01	0,22	48
Cl^-	0,28	0,83	0,01	0,23	52

The course of the suspended dust concentrations described in fig. 9, shows the highest aerosol concentrations for sulphate with 2.3 µg/m³ in July 1983, the lowest concentrations were recorded in winter, especially in December 1983 with 0.35 µg/m³. Nitrate showed high monthly averages in March 1983 with 0.61 µg/m³, June 1983 (0.55 µg/m³), March 1984 (0.54 µg/m³), and September (0.50 µg/m³). The lowest concentrations were found in October 1983 (0.04 µg/m³) and December 1983 (0.08 µg/m³).

The monthly average value for chloride varied between 0.56 µg/m³ October 1984) and 0.06 µg/m³ (February 1984).

2.3.3 <u>Comparison of the total deposition ('wet and dry-only') between polluted and non polluted area</u>. Table 5 contains the complete dry plus wet (however, not the gaseous one) deposition of sulphate, nitrate and chloride particles for the station in the non polluted high-

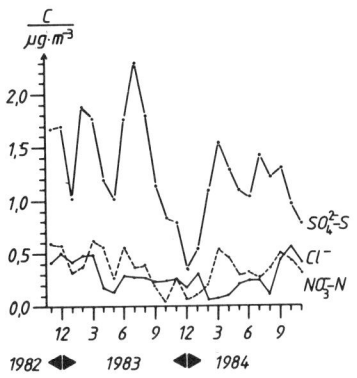

Fig. 9 Monthly average of aerosol-concentration at the mountain station in the non polluted area (11/1982 - 11/1984)

Tab. 5. Total deposition and percentage of dry deposition at the station in the industrial area (Bochum) and at the mountain station (Kahler Asten (in mg/m²· d) (1/1983 - 2/1984)

	SO_4^{2-} - S			NO_3^- - N			Cl^-		
	$\bar{x}$	$\bar{x}$(dry) (%)	rel.dev. (dry) (%)	$\bar{x}$	x(dry) (%)	rel.dev. (dry) (%)	$\bar{x}$	$\bar{x}$(dry) (%)	rel.dev. (dry) (%)
mountain station	7,26	21,8	9,4	2,35	20,3	11,8	6,18	15,5	12,9
industrial area	6,60	34,1	16,3	1,93	26,8	18,2	5,14	29,2	15,5
m.s. - i.a.	0,66	-12,3	-6,9	0,42	-6,5	-6,4	1,04	-13,7	-2,6

land area as well as for the station in the industrial area.

The mountain station shows values which are for sulphate 10 % higher than the rates at the station in the industrial area. For nitrate the rates are 22 % higher, for chloride 20 % higher. All values apply to the total deposition.

While at the mountain station the shares of dry deposition are 15.5 % and 21.8 %, these rates are distinctly higher in the polluted area, where the share of dry deposition is 26.8 % to 34.1 %.

Besides the fact that the dry deposition in the non polluted region plays a less important part, the standard deviation of the dry depositions compared to their average values show that the occurrence of the above mentioned suspended particles in that area is subject to less deviations than in the polluted area.

3. DISCUSSION OF THE RESULTS

The Sauerland region, the remote highland area under investigation, receives pollutants predominantly from the long-distance transport, with the exception of some small industrial sources in the valleys.

3.1 'bulk'-precipitation values and immission rates

The computed regional average values of the 'bulk'-concentrations fit into the scheme of other rural regions in the FRG (GEORGII et al. 1980).

The spatial variability of the sulphate, nitrate and chloride concentrations was nearly 20 % apply to the average. The fluoride value was higher due to an industrial plant in the area.

The average pH-values show only insignificant deviations for the acid end of the scale, but greater deviations (up to 3 pH-units) for the basic end. Similarly high amplitudes are reported by GEORGII et al. (1982).

Essentially greater spatial and temporal differences were noticed for the immission rates (SAM-stations).

Besides the greater variability of the sulphur-immission rates in the Sauerland in winter the absolute rates in the cold season are twice as high as in summer. The reasons for this are on one hand the increasing demand for heat and the resulting higher SO_2-emissions near the ground due to the dropping temperatures, and on the other hand the increasing occurence of weather situations with almost low exchange conditions.

Since the immission rate is made up of the immission concentration and the run of the wind it cannot be decided, whether high or low concentrations or low respectively high wind speeds are responsible for the immission rate determined by the measuring station, without measuring the wind speed.

To my opinion it is not permissible to conclude from the immission rate to the immission concentration, as SCHUH (1985) did.

3.2 'wet and dry'-only values

The dry and wet deposition measured in the industrial area and in the highland area seperately correspond to those obtained in other slightly loaded areas of the FRG (PERSEKE 1982).

The most distinct annual curve is shown for the chloride concentrations with maximum values in winter and low rates in summer. One reason for the higher chloride concentrations in winter might be the frequent occurence of rain-bringing west winds from the sea; the partly occuring summer maxima of the sulphate concentrations might be explained by the higher photochemical production rate during the summer months, which produces more sulphur in this period.

The greatest dependence of the concentration of chemical constituents in precipitation on the wind direction in the non polluted region was shown by chloride with high concentrations in the case of precipitation from northwest; for continental precipitation from the east, however, the lowest concentration was recorded. Since either the North Sea and the Ruhr District are situated in the westnorthwest of the mountain station it cannot be cleared up exactly, where this increase in concentration can be put down to.

As the examinations at the industrial area station for chloride, dependent on air-massis and wind direction, also show a clear increase in the concentration from the northwest sector (results of an earlier unpublished paper of the author), - and thus show that they are influenced by the sea - this may also be true for the mountain station.

Comparative measurements concerning the problem of 'rain-out' and 'wash-out' between industrial area station and mountain station (845 m above sea-level) revealed that the precipitation sum measured in the industrial area was 56 % of that registered at the mountain station, with the share of sulphate being 77 %, of nitrate being 71 % and of ammonium as well as chloride being 91 % of the share at the mountain station. Compared to the lower amount of precipitation in the industrial area higher inputs were recorded.

Supposing that the greatest share of the wet deposition at the slightly loaded mountain station mainly results from the 'rain-out', the deposition measured at the station in the industrial area (140 m above sea-level) result from the 'wash-out' of the 700 m thick airmass between the mountain station in the non polluted region and the station in the industrial area. After various measurements in Frankfurt/M. and on the small Feldberg/Taunus (GEORGII 1965) similar results were obtained, with the exception of the slightly deviating rate for sulphate; the same result was obtained for the 'wash-out' shares in the Soviet Union with 45 % to 78 % (PETRENCHUK & SELEZNEVA 1970).

From the data available (aerosol concentration and deposition rates) the deposition velocity for sulphate (v_d = 1.4 cm/sec), nitrate (v_d = 1.6 cm/sec.) and for chloride (v_d = 4.5 cm/sec) were computed.

The deposition rates of nitrate and chloride - the rate for sulphate corresponds to that measured by EVRETT et al. (cit. in DAVIES & NICHOLSON 1982) - are higher than those determined in the laboratory (compilation of individual values in PERSEKE 1982).

Concerning the question of the percentage of the most important acid-forming anions in the precipitation it can be stated that at the measuring station in the polluted area and at the station in the non polluted area more than 50 % fall to sulphate ions; at the station in the industrial area chloride ions rank second with 26 %, then follow the nitrate ions which make 19 %. At the mountain station, however, nitrate is second before chloride; responsible for the higher share of chloride in the industrial area are probably the shorter distance between the sampling station and the sea, but also the release of HCl caused by industrial processes.

ACKNOWLEDGEMENTS

It would not have been possible to do this work without the generous financial and technical help by the Institute of Meteorology and Geophysics (Head: Prof. Dr. H.-W. Georgii) and the Umweltbundesamt, pilot station, both Frankfurt/M.

References

DAVIES, T.D. & K.W. NICHOLSON (1982): 'Dry Deposition Velocities of Aerosol Sulphate in Rural Eastern England', pp. 31 - 42. - In: GEORGII, H.W. & J. PANKRATH (eds.)(1982): Deposition of Atmospheric Pollutants.

FUKUI, K. (1966): 'Die alkalische Filterpapiermethode für die Bestimmung der Schwefeloxyde und auch der Stickstoffdioxyde und Chloride in der Luft'. - In: Proc. of Int. Clean Air Congress, London, 4. - 7. Oct. 1966, p. 231 f.

GEORGII, H.W. (1965): 'Untersuchung über Ausregnen und Auswaschen atmosphärischer Spurenstoffe durch Wolken und Niederschlag'. - In: Ber. Dtsch. Wetterdst. 14, Bd. 100.

GEORGII, H.W.; PERSEKE, C.; ROHBOCK, E. & G. GRAVENHORST (1980): Untersuchung über die trockene und feuchte Deposition von Luftverunreinigungen in der Bundesrepublik Deutschland'. Umweltforschungsplan des Bundesministers des Inneren. Forschungsprojekt 10402600, Inst. für Met. u. Geophy., Frankfurt/M..

GEORGII, H.W.; PERSEKE, C. & E. ROHBOCK (1982): 'Feststellung der Deposition von sauren und langzeitwirksamen Spurenstoffen aus Belastungsgebieten'. - Umweltforschungsplan des Bundesministers des Inneren. Luftreinhaltung, Forschungsprojekt 10402600 sowie Datenband. Im Auftrag des Umweltbundesamtes.

KUTTLER, W. (1985): 'Einträge atmosphärischer Spurenstoffe in Waldgebiete des Sauerlandes'. - Umweltforschungsplan des Bundesministers des Inneren; Luftreinhaltung, Forschungsprojekt 10402715. Im Auftrag des Inst. f. Met. u. Geophys., Univ. Frankfurt/M. u. d. Umweltbundesamtes, Pilotstation Frankfurt/M..

PERSEKE, C. (1982): 'Die trockene und feuchte Deposition säurebildender atmosphärischer Spurenelemente'. - In: Ber. Inst. Met. u. Geophys. Univ. Frankfurt/M., Nr. 48.

PETRENCHUK, O.P. & E. SELEZNEVA (1970): 'Chemical composition of precipitation in regions of the Soviet Union'. - In: Geophys. Res. 75, pp. 3629 - 3634.

RUMPEL, K.J. (o.J.): 'Ein Verfahren zur Feststellung flächendeckender Immissionsraten mit dem Immissionsratenmeßgerät SAM nach RUMPEL'. Unveröff. Manuskript.

SCHUH, A. (1985): 'Luftqualitätsbestimmung in Kurorten'. - In: Annalen d. Meteorologie (N.F.), Nr. 22, pp. 88 - 89.

SMITH, F.B. (1983): 'Conditions pertaining to high acid concentrations in rain'. - In: VDI-Berichte, Nr. 500, pp. 67 - 75.

WINTER DEPOSITION RATES OF ATMOSPHERIC TRACE CONSTITUENTS IN FORESTS: ASSESSMENT OF TOTAL INPUT.

G. Glatzel, M. Kazda and G. Markart
Institute of Forest Ecology
Universität für Bodenkultur
Peter Jordanstrasse 82
A-1190 Wien, Austria

ABSTRACT: During winter, the assessment of total input of atmospheric trace constituents into forest ecosystems is extremely difficult. Besides deposition with rain and snowfall, contributions of complex dry deposition processes have to be accounted. Aerosol deposition to physically and chemically changing snow surfaces on the ground and in the forest canopy plays an important role. Hoarfrost formation, riming or impaction of non freezing fog droplets on tree crowns, and resuspension of heavily polluted surface snow during storms and subsequent deposition in snow drifts may be very important processes in certain areas. Results of winterly deposition studies in forest ecosystems of the Vienna Woods and of the Warscheneck mountains in Upper Austria are used to exemplify some of the problems. A snow lysimeter technique, designed to collect meltwater from the snowpack, is being currently tested as an integrative method for estimating total winterly input.

1. INTRODUCTION

For many forest ecosystems, deposition of atmospheric trace constituents has become the most important source of biogeochemically effective substances. Unfortunately the methods available to assess total annual input to forest ecosystems are still insufficient. Besides problems associated with measuring deposition rates of trace gases to the tree canopy, estimation of winterly deposition rates is still rather inaccurate for regions with winter frosts and snowcovers. As winterly deposition may contribute considerably to total annual deposition rates, even in areas with rather moderate winter conditions (Glatzel et al. 1986), methods to measure winterly deposition should be improved.
During winter the amounts of trace constituents deposited on forest ecosystems with rain and snow may be modified considerably by the following processes.

1.1. Dry deposition to the snow surface

As freshly fallen snow ages, it becomes enriched with various substances. As far back as 1892, Carter Bell observed an increase in acidity of the snow in the vicinity of a chemical industry. Since then, deposition of aerosols to snow has become a well studied phenomenon (Dovland and Eliassen 1976, 1977; Brimblecombe et al. 1985; Cadle et al. 1985). Deposition flux density depends not only on air quality and temperatures, but also on chemical and physical properties of the snowpack.

Much less is known of the role of hoarfrost-formation on the snow surface as a mechanism of pollutant deposition. The surface of the snowpack may reach extremely low temperatures during clear nights, establishing steep diffusion gradients for water vapour. Some pollutants appear to be cotransported to the growing ice crystals. In alpine valleys with local sources of air pollutants, hoarfrost precipitation out of valley-mist and fog, during long periods of winterly temperature-inversions, seems to be a major deposition mechanism.

Resuspension of polluted surface snow by storms and subsequent deposition in snow drifts can be an effective mechanism of local pollutant enrichment. In deciduous forests, where strong winterly winds meet little resistance in the canopy, redistribution of snow is quite important. Plant species, which indicate high nitrate levels in the soil, have recently become much more common in the deposition zone of snowdrifts leeward of ridges in the Vienna Woods (Glatzel et al. 1986). It is likely that excessive nitrogen input from accumulations of polluted surface-snow in the snowdrifts contributes to this phenomenon.

1.2. Dry deposition to the canopy

Deposits of snow, rime and hoarfrost on branches may substantially increase the surface area of the canopy available for dry deposition. In dense stands hoarfrost forms mainly in the upper parts of the canopy where greater turbulence lowers boundary layer resistance for diffusion processes.

Canopy surface riming by impaction of supercooled, polluted fog droplets can play an important role on exposed ridges in areas where winter fogs are frequent (Borys et al. 1983).

2. EXPERIMENTS

2.1. Effects of riming

Fig. 1 shows snowpack quality in mature beech stands along a profile of nearly 15 km length through the Vienna Woods from the Tullnerfeld in the West, to the city of Vienna in the East. During an extreme cold spell in January 1985, the aging of the snowpack could be monitored without disturbances by snowfall or elution due to thawing. Chemical analyses showed that topography was highly correlated with pollution of the snow surfaces. Extreme deposition on hill tops and ridges could be traced to riming from low clouds or fogbanks brushing the exposed tree crowns. Eventually rime-growth falls off from the branches and collects on the surface of the snowpack.

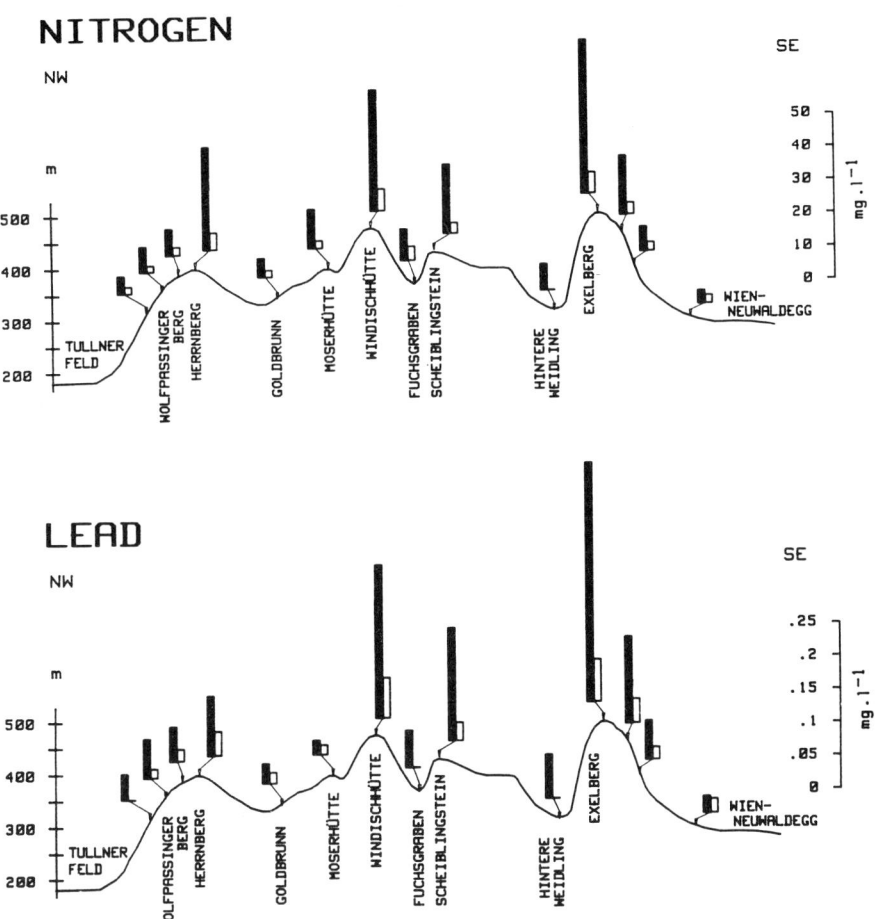

Fig.1 Deposition of atmospheric trace constituents to the snowpack in beech stands along a profile through the Vienna Woods. Exposed ridges show extreme depostition due to canopy riming.
a) Concentration of total dissolved nitrogen in the snow-water
b) Concentration of lead in the snow-water
Solid bars: top 2 cm of snowpack
Open bars: total snowpack
Elevations tenfold exaggerated

At temperatures above freezing, impaction of fog droplets on branches and subsequent stemflow and crowndrip are important. Kazda

(1985) has shown, that fog interception by tree crowns is a dominant process of pollutant deposition on exposed hilltops of the Vienna Woods. Results of soil analyses in 160 beech stands of the Vienna woods indicated an increase of lead content in the topsoil from the valleys to the hilltops (Glatzel et al. 1986), emphasizing the ecological significance of these deposition mechanisms.

2.2. Deposition to the snow surface

Fig. 2 shows deposition flux densities for total dissolved nitrogen and sulfate to the snow surface in spruce stands along a transect from 900 to 1500 metres elevation at Wurzeralm, Upper Austria.

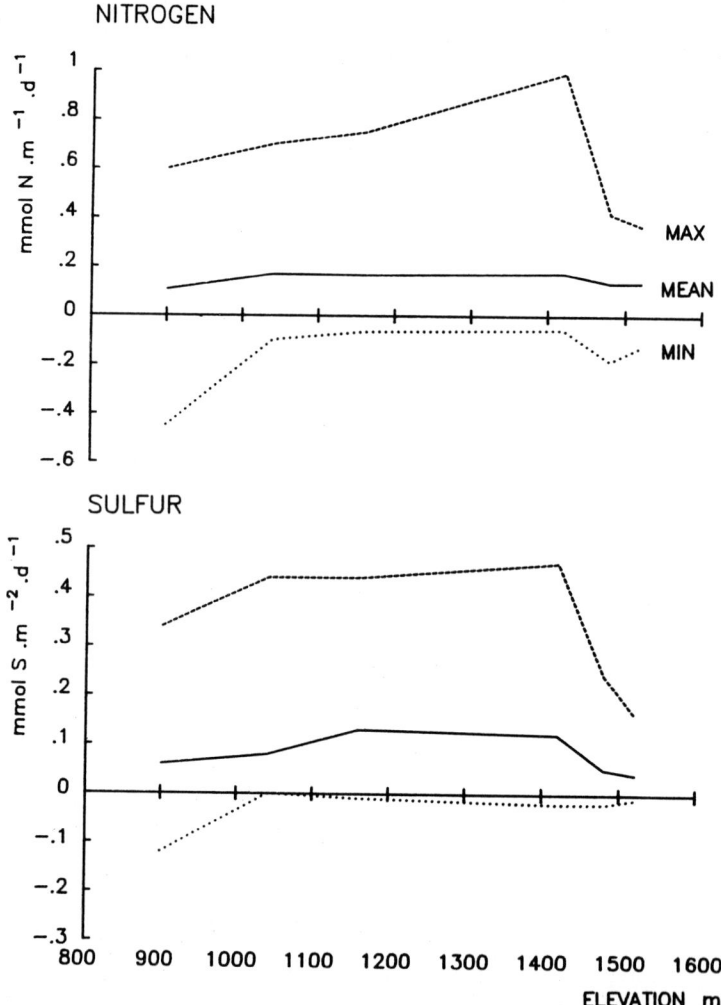

Fig.2 Deposition flux densities of nitrogen and sulfur to the snowpack in spruce stands at the Wurzeralm transect in Austria, during periods without snowfall or rain.

For this investigation the snowpack was sampled daily with a snow corer down to reference surfaces in the snowpack. The reference surfaces were PE-sheets and later PE-meshware, allowed to be covered by snowfall. To calculate the values for fig.2, days without snowfall and with subzero temperatures from a three-month-period (January to March 1985) were used. The data show large variation between minimum and maximum values. Negative values resulted from resuspension of snow by wind, extreme positive values from deposition of resuspended snow and from snow, hoar and rime blown off from tree crowns. The comparatively large negative deposition rates observed at the lower profile point might also result from ion leaching in the snowpack at temperatures close to zero C.

In the investigated spruce stands annual secondary deposition to the snowpack, may be estimated to be 5 to 10 kg $N.ha^{-1}$ and 12 to 18 kg $S.ha^{-1}$, excluding primary input by snowfall or rain. This is a considerable contribution for an area not affected by local pollution sources.

3. METHODS FOR ESTIMATING TOTAL WINTERLY INPUT INTO FOREST ECOSYSTEMS

So far, no simple method is available to estimate the total input of atmospheric trace constituents into forest ecosystems during winter. The following discussion of various approaches is based on the experience we gathered from deposition studies in Austria.

3.1. Heated rain gauges, wet-only samplers and snow buckets

With these instruments only rain and strong canopy drip can be collected with sufficient accuracy. For snow the correlation of the data obtained from these devices with data from core sampling is frequently poor, especially under conditions of windy and stormy weather. Gauges placed above snow level underestimate deposition, snow buckets set into the snowpack are easily blown full of snow. Additional problems arise with wet-only samplers, as their heated rain sensors are frequently not triggered by fine light snow during cold weather. Dry deposition of aerosols to the snowpack and secondary redistribution of coarse snow crystals, swept just above the snow surface by strong winds, is neglected. Snow, ice and rime falling from the canopy are difficult to collect as these deposits show extreme spatial variation and require a large number of collectors in forest stands.

3.2. Snow profiles

Using coring tools, sampling of snow profiles is simple and a sufficient number of replications can be collected at reasonable cost. While this method is excellent for estimating water content of a snowpack, its use for assessing deposition of trace constituents is very limited.

Except at very low temperatures, solutes may migrate in snow and thawing of surface snow or rainfall may lead to rapid elution of salt- and acid-ions (Cadle et al. 1984, Davies et al. 1982, Johannessen and Henriksen, 1977). Small amounts of percolating water are sufficient to leach out large amounts of solutes. Periods of thaw and rainfall tend to interrupt the winter cold in our climate. Besides this, topography and stand structure influence thaw processes. For these reasons, the use of snow profile methods to investigate regional deposition patterns, must be viewed very critical.

3.3. Daily snow core sampling

For this method snow cores are collected daily. In areas with deep snow covers PE-meshware may be placed on the snow surface every few weeks and allowed to be covered by new snow, providing a reference for sampling depth. Snow buckets or rain gauges must be installed to collect rain and very wet snow.

This method is quite accurate and provides valuable information on deposition processes to the snowpack. Under special conditions interpretation is difficult, e.g. if a storm erodes the snow surface and rain leaches the snow before next days measurements. The most serious drawback of this method is its high cost. One person is needed in the field throughout the winter and the large number of samples entails high analytical expenses.

3.4. Snow lysimeters

In search for a simple method to assess total winter in- put of trace substances into forest ecosystems, we tried the following procedure.

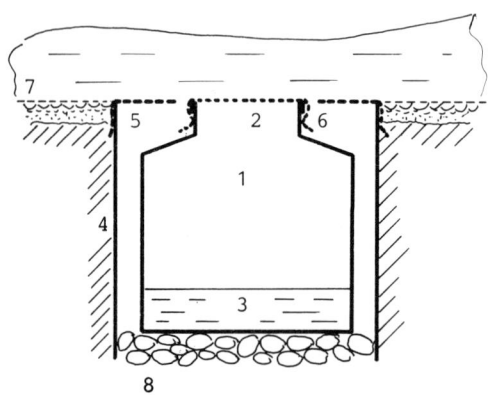

1 PE vessel, 20 cm wide-mouth
2 PE meshwork, 2 mm
3 collected water, spiked with mercury acetate to inhibit microbial activity
4 section of PVC tubing (support)
5 stiff polyamide meshwork
6 gap to trap migrating soil fauna
7 fine PP-gauze to prevent splashing of soil particles
8 gravel or styrofoam support

The dug out soil has to be carefully collected on plastic sheets and removed. The litter layer around the sampler must not be disturbed.

Fig.3 Snow lysimeter for collecting total winter deposition

Wide-mouth containers with an opening of 20 to 30 cm diameter and
enough capacity to collect all winter precipitation were placed in
holes dug into the forest soil. The openings were covered with 2 mm
PE-meshwork. The first snowfall blankets such collectors and a snow-
depth of 10 to 20 cm is sufficient to make them indistinguishable in
the snowcover. In order to avoid trapping game or tourists, a fence is
mandatory. Meltwater percolating through the snow is collected in
these containers. These lysimeters gave reasonable values for total
precipitation. With regard to trace constituents, problems arose from
growth of algae on the container walls and from contamination by spla-
shed soil material and by entrapped soil fauna.

An improved version (Fig. 3) is currently being tested and we hope
that this method can eventually be developed to a point where it can
provide a fair estimate of total winter deposition at reasonable cost.

ACKNOWLEDGEMENT: This investigation was supported by the Austrian
Ministry for Science and Research, Project Nr. 36.006/1-23/83 and
36.017/1-23/84

REFERENCES

BORYS, R.D., P.J. DEMOTT, E.E. HINDMANN and D. FENG (1983) The signi-
    ficance of snow crystal and mountain-surface riming to the re-
    moval of atmospheric trace constituents from cold clouds. In:
    H.R. PRUPPACHER, R.G. SEMONIN and W.C.N. SLINN (Eds.): Precipi-
    tation scavenging, dry deposition and resuspension. P. 181-189,
    Elsevier, New York
BRIMBLECOMBE, P., M. TRANTER, P.W. ABRAHAMS, I. BLACKWOOD, T.D. DAVIES
    and C.E. VINCENT (1985) Relocation and preferential elution of
    acidic solute through the snowpack of a small, remote, high-
    altitude Scottish catchment. Annals Glaciology 7: 141-147
CADLE S.H., J. MUHLBAIER DASCH and N.E. GROSSNICKLE (1984) Northern
    Michigan snowpack - a study of acid stability and release.
    Atmospheric Environment 18: 807-816
CADLE, S.H., J. MUHLBAIER DASCH and P.A. MULAWA (1985) Atmospheric
    concentrations and the deposition velocity to snow of nitric
    acid, sulfur dioxide, and various particulate species. Atmos-
    pheric Environment 19: 1819-1927
CARTER BELL J. (1892) Analysis of snow around a chemical industry. J.
    Soc. Chem. Ind. 11: 320
DAVIES T.D., C.E. VINCENT and P. BRIMBLECOMBE (1982) Preferential elu-
    tion of strong acids from a Norwegian ice cap. Nature 300: 161-
    163
DOVLAND H. and A. ELIASSEN (1976) Dry deposition on a snow surface.
    Atmospheric Environment 10: 783-785
DOVLAND H. and A. ELIASSEN (1977) Estimates of dry deposition on snow.
    IR 34/77 SNSF, Oslo

GLATZEL, G., M. KAZDA und L. LINDEBNER (1986) Die Belastung von Buchenwaldökosystemen durch Schadstoffdeposition im Nahbereich städtischer Ballungsgebiete: Untersuchungen im Wienerwald. Düsseldorfer Geobot. Kolloq. 3: 15-32

JOHANNESSEN, M. and A. HENRIKSEN (1977) Chemistry of snowmelt water: changes in concentration during melting. FR 11/77 SNSF, Oslo

KAZDA, M. (1985) Untersuchung von Schwermetalldepositionsvorgängen aus Analysen fraktionell gesammelter Stammabflußproben und Jahresgang der Schwermetalldeposition in einem Buchenwaldökosystem des stadtnahen Wienerwaldes. Diss. Universität für Bodenkultur, Wien

Interception
of Fogwater

DESIGN OF A FOG WATER COLLECTOR FOR CHEMICAL ANALYSIS

Frank Dröscher
Universität Stuttgart
Institut für Verfahrenstechnik und Dampfkesselwesen
Pfaffenwaldring 23, D 7000 Stuttgart 80

ABSTRACT. Design requirements for fog water collectors for chemical analysis are derived from physical and chemical properties and the principles of deposition of fog droplets. Advantages of different collector types are discussed. The design of a new high-volume string collector with high capturing efficiency, defined moderate impaction conditions, and the potential for on-line chemical analysis is presented along with the results obtained in first field tests.

1. INTRODUCTION

In recent years the collection of fog water for chemical analysis has been the subject of increased scientific activity. Major applications of fog water collection as an essential tool in experimental investigation are cloud chemistry, especially the heterogeneous chemistry of atmospheric pollutants, and problems arising from the deposition of fog or cloud water on materials and vegetation with the effects of corrosion and plant damage. In the light of the extensive forest die-back observed in many regions of Central Europe and other parts of the industrialized world the contribution of fog or cloud water deposition to the nutrient and pollutant input into forest eco-systems has gained considerable interest.

2. FOG WATER DEPOSITION

Basically there are two principle approaches to the assessment of fog water depositon. The first one calculates the contribution of fog water depositon from the difference between the total wet deposition including the fog water deposition and the precipitation deposition. In case a forest canopy itself serves as droplet impactor bulk collectors distributed within and outside the forest stand provide the samples, usually on a weekly or biweekly basis, at best on an event basis /1/. Similarly, screens or nets on top of bulk samplers have been employed as fog droplet impactors /2,3/. Depending on the exposition interval, however, the resulting samples reflect not only fog water deposition but also the effects of dew formation and dry deposition of gases and particles. Interception, i.e. wet deposition on vegetation that evaporates before reaching the ground, as well as concentration peaks of individual fog water components, which both lead to high pollutant concentrations in the wet film covering leaves and needles, cannot be detected by this method. Pollutant peaks, however, may exert more extreme plant demage than average concentrations.

The second approach to assess the contribution of fog water to the chemical input by wet deposition combines the knowledge of the chemical composition of fog droplets with a physical model of fog droplet deposition. Since the impaction of droplets on vegetation is a complex function of atmospheric turbulence, advection, surface properties, geometry etc. a parameterized general solution of the impaction process on the scale of forest stands seems impossible to obtain. Empirical solutions with limited special and temperal validity, however, may be attained. This way the physical and chemical analysis of fog droplets is a first step to assess the chemical input via fog water deposition.

The broad spectrum of scientific questions and approaches has led to the development of a large number of fog water collectors devised for different purposes. In this paper, design criteria for fog water collectors in general are discussed on the basis of the physical properties and the design of a new high-volume fog water collector is presented along with preliminary field data.

2. PHYSICAL PROPERTIES OF FOG

Fog forms near the ground, in the same manner as clouds in the general atmosphere, upon supersaturation of wet air by condensation of water vapor on non-activated nuclei: either on dry particles or, at relative humidities 70 % to 100 %, on wetted aerosol. In the presence of hygroscopic nuclei fog formation can start at relative humidities as low as 70 %.

These non-activated nuclei are predominantly on the order of 0.01 to 1 µm in diameter, separated into two distinct size modes /4/. Activated nuclei grow to drop sizes well above 1 µm. Due to increased gravitational forces droplets bigger than 60 µm to 100 µm in diameter, depending on the degree of turbulence do not remain suspended in the atmosphere. Fog droplets therefore range between $d_T \approx 5$ µm and $d_T \approx 60$ µm (100 µm) in size. They may coexist with interstitial dry aerosol ($d_T \approx 1$ µm) and rain or snow ($d_T \approx 60$ µm (100 µm)).

Since fogs differ in their meteorological development fog types with different properties can be distinguished. Typical values and average range of main fog parameters were given by Jiusto /5/ for two major fog types (table I):
- radiation fog due to noctural radiational cooling of underlying surfaces usually of local extent
- advection fog associated with air mass transport and consequently with wind and of regional extent.

TABLE I: Typical values of fog properties cited from Jiusto/5/

Parameter	Radiation Fog	Advection Fog
Droplet diameter	$5...10...35$ μm	$7...20...65$ μm
Liquid water content	110 mg/m^3	170 mg/m^3
Droplet number conc.	200 $/cm^3$	40 $/cm^3$
Vertical fog depth	100 m, max. 300 m	200 m, max. 600 m
Horizontal visibility	100 m	600 m

The liquid water content in fog usually makes up 1% or less of the total water content of fog air, hardly exceeding 2% to 4%. Small alterations of the thermodynamic state of fog air given by its pressure and temperature results in drastic changes of the liquid water content, e.g. in the partial or total disappearance of the fog /6/. Due to this delicate thermodynamic balance, fog properties are often subject to huge special and temporal variations even in the course of a single fog event. This pertains particularly to fogs over complex terrain.

More detailed information on fog properties is provided in fog size spectra obtained by optical drop size analysis methods. As an example, in figure 1 the discrete number distribution $d\,n(d_T)/d\,d_T$ and the corresponding relative liquid water content w_L were plotted as function of the aerosol diameter d_T for three typical fog size spectra. Kunkel /7/ recorded these spectra at the Atlantic coast of Massachusetts/USA. In principle these distributions agree well with those measured in other fog types, e.g. radiations fog /8/.

In fog non-activated nuclei of diameter $d_T \leq 2$ μm are typically found to constitute the majority of aerosols. A strong decline in number is seen between $d_T \approx 2$ μm and $d_T \approx 5$ μm which is in agreement with thermodynamics of condensation and scavenging mechanisms. Within the size range of fully developed droplets, a second maximum appears at $d_T \approx 10$ μm (type B and C) or $d_T \approx 30$ μm (type A) in all three spectra presented in figure 1a). The plateau in type B around $d_T = 25$ μm can be explained as the result of superposition of type A and C and was found to develop into a distinct third maximum in other fog events /8/.

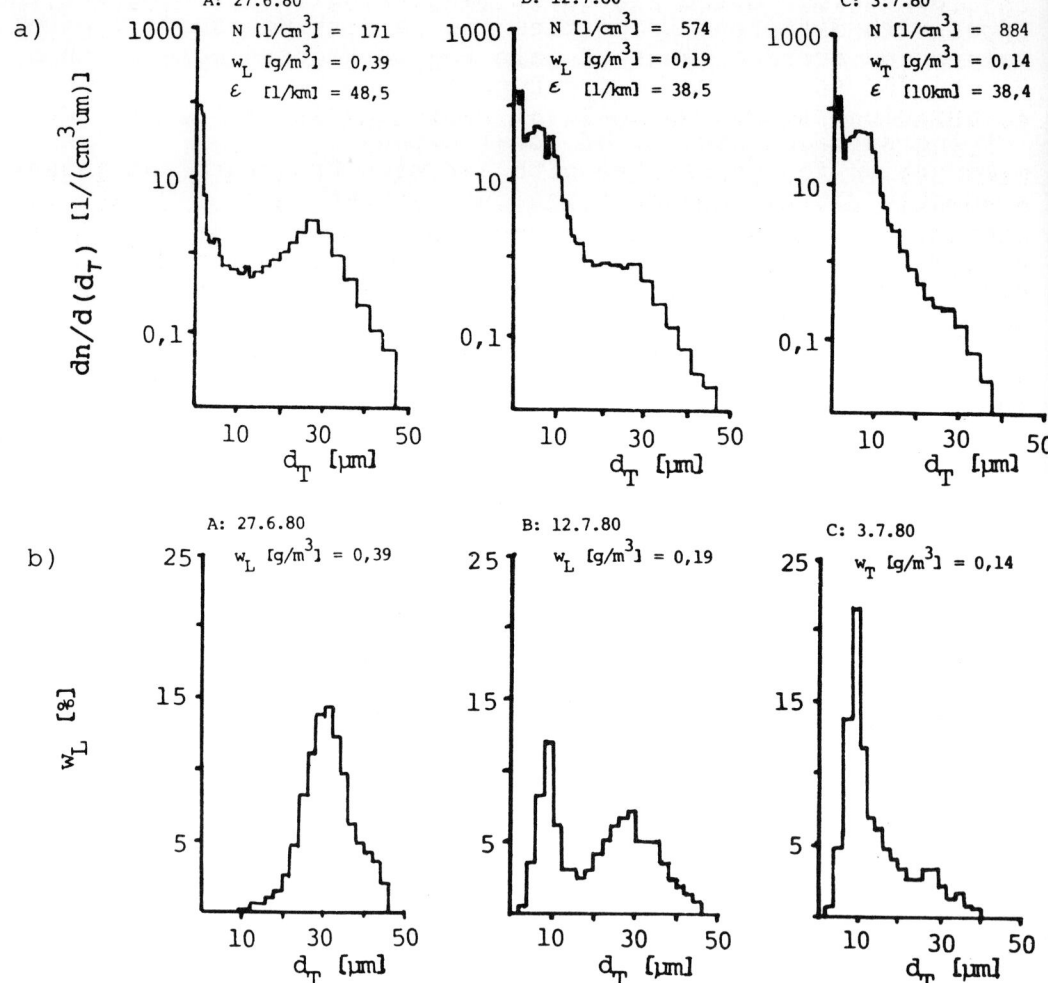

Figure 1. a) Discrete aerosol number distribution $d\,n(d_T)/d\,d_T$
b) Relative distributon of the liquid water content w_L
as functions of the aerosol diameter d_T for 3 types of fog size spectra /7/

Looking at the distribution of the liquid water content, one realizes that the contribution of droplets smaller than 5 μm (type B and C), in some cases those smaller 20 μm (type A), can be neglected. With respect to the contribution of small droplets to the total solute mass in fog water, these lower limits would have to be set somewhat lower, because of the presumably higher solute concentrátions in small droplets.
In addition to the physical properties of fog, the possibility of chemical transformations in the water sample during

the **collection period has to be considered in the** design of a fog water collector.

4. CHEMICAL TRANSFORMATIONS IN THE FOG WATER SAMPLE

With respect to its chemical composition, fog water represents a complex mixture of electrolytes with undissolved fractions similar to rainwater. Due to the dynamics of condensation and evaporation, the varying composition of the surrounding gas phase, and on-going slow reactions within the droplets, fog water is not in chemical equilibrium.

During the time required to collect the minimum sample volume, which usually is 10 ml to 50 ml for an analysis covering all species contributing to the ion balance, transformations of the chemical composition can occur.

In the course of the sampling process itself, interactions of fog water with the surrounding air may lead to:
- concentration of dilution of fog droplets due to changes in the saturation of the fog air as it approaches the impactor unit of a collector, or of droplets already deposited in the impactor unit as a result of variations of the fog composition or due to frictional heating in impactors working with high pressure drops
- absorption in or desorption of water soluble gases from water droplets deposited in the impaction unit
- depositon of particles especially on wet impactor surfaces
- collection of wet deposition other than fog, e.g. dew or rain.

Interactions of sampled water with collector surfaces have to be considered, too. Particularly,
- absorption of fog water components at collector surfaces
- leaching of trace elements from the construction materials
- chemical transformations within the sample resulting from the activity of bacteria and algae growing attached to surfaces.

Finally, reactions of fog water components themselves occur due to suspended bacteria and deviations from the chemical equilibrium. Depending on the temperature and oxidation conditions in the water sample, rapid concentration changes for less stable fog water constituents such as H^+, NH_4^+, NO_3^-, and H_2O_2 may be observed even during a half hour interval /9/.

Based on these principle physical and chemical properties of fog and fog water relevant to fog water collection, in the subsequent section design criteria for fog water collectors will be developed.

5. DESIGN CRITERIA FOR FOG WATER COLLECTORS

A large number of fog water collector designs has been reported in literature. Generally, inertial impaction is used for the deposition of fog droplets . It is the only effective separation method that does not affect the fog water composition.

Passive samplers rely on the natural wind as driving force for the deposition of fog water in a stationary impaction unit usually a screen of some kind.

Active samplers accelerate the fog air relative to the impactor unit. There are rotating collectors which provide a relative velocity between the fog air and the impaction unit by the rotational movement of the impactor either of rods or a string harp. The impacted water is driven to the sampling vessel by centrifugal forces. Vacuum collectors use a vacuum pump to suck fog air through a duct that houses the impactor unit. Vacuum collectors fall into three groups: vacuum samplers with screen impactors comprised of nets, filter mashes, or string harps, cyclones which force the fog air into a rotational flow field thereby precipitating the droplets to the cylindrical cyclone wall, and jet impactors that accelerate the fog air towards an impaction plate or a stagnant air buffer where droplets accumulate because they cannot follow the sharp bend of the streamlines in front of the obstacle. All three types depend on gravity or a special pump for the transfer of sample water to the collection vessel.

Any of these fog water collector has to fulfill the basic functions of:
- sampling fog air
- impaction of fog droplets in the collectors impaction unit
- transfer of the deposited fog water to a sampling unit.

With respect to sampling of fog air, the principle objective is to deposit fog droplets from fog air of representative composition. Care has to be taken to allow the entire fog size spectrum to reach the impaction unit. This may be critical for the large fog droplet fraction ($d_T \gtrsim 60 \mu m$) because these fail to follow sharp bends in the air flow. Therefore collectors with preferential sampling direction should be directed towards the wind. Active collectors should operate at a near isokinetic air sampling rate. The possibility of misrepresentation of large droplets should influence the design and direction of air inlets of vacuum collectors and the choice of the axis of rotation in which fog air approaches a rotating collector.

In order to distinguish between fog water and precipitation water, the impaction unit and the collection vessel should be protected against the deposition of rain and snow.

Backmixing of air from down-stream the impaction unit with untreated air in front of the impactor also results in a shifted fog composition of the sampled air. While multiple passages of fog air through the impactor unit increases its **impaction** efficiency, short circuiting of air from the outlet of a vacuum collector after frictional heating in the pump, or within a rotating collector is to be ruled out. Only then the liquid water content of the fog can be estimaed from the sample water volume.

Sufficient distance from the ground and objects should be ensured to avoid fog water depletion in the sampled air resulting from droplet deposition on vegetation and objects due to increased turbulence. Heat transfer from objects (e.g. motors) or the ground influences the saturation condition of fog air in their vicinity, leading to partial or total evaporation of fog droplets, or, only in case of cold surfaces, to condensation.

Since the sampling period necessary to collect a set minimum sample water volume for the chemical analysis depends on the liquid water content of the fog, the impactor effciency, and the rate of fog air sampling, the air sampling rate is a decisive design parameter. High air sampling rates lead to a good time resolution of the measurement as well as they minimize the possibilility of chemical transformation of the water sample.

The impaction of fog droplets in the impactor unit constitutes the core function of the collector. Here liquid water is separated from the gas phase. The efficiency of this separation is a function of the droplet size as shown later. In order to sample fog droplets representatively the entire size spectrum of fog droplets (5 µm $\lesssim d_T \lesssim$ 60 µm (100 µm)) should be deposited with a constant high efficiency excuding the smaller non-activated particles. This can only be achieved by defined constant impaction conditions which is not given for passive collectors. When certain size fractions are to be sampled a sharp size cut at both ends of the size range has to be realized. In any way the impaction characteristics of the collector should be investigated.

Finally, fog water deposited in the impactor unit has to be transferred to a collection bottle or, possibly, to a continuously working analyser for characteristics, like pH or conductivity, or chemical species.

In order to minimize interactions of the sampled water with the surrounding air and construction materials as well as homogeneous liquid phase reactions and evaporation losses the residence time of the sample water in the impactor unit and on the migration path to the collection bottle should be short. This ought to be considered when designing the sampling rate and the migration path.

The wetted surface area should be kept small and chemically inert surface materials be selected. While for the analysis of inorganic water components polyethylen (PE), polypropylen (PP), and teflon (PTFE) perform satisfactorily, glass is preferred for the analysis of organic components. Nylon should be avoided because it tends to adsorb ammonia preferentially.

Besides storing the sample water, the collection bottle should also serve as effective protection against evaporation losses and any contaminations. During fog free periods the entire impaction unit and the sample water migration path should be protected from both dry deposition of gases and particles as well as precipitation and dew formation.

For convenient cleaning of these collectors parts before the sampling period all parts should be easily accessible and narrow slots be avoided. Continuous sample water flow allows for immediate or on-line analysis of instable water constituents.

These general design criteria express the principle requirements any fog water collector should fulfill. Concrete design parameters for fog water collectors have to be developed from the inertial impaction theory which provides impaction parameters.

6. IMPACTION PARAMETERS

For the design of fog water collectors inertial impaction theory allows to determine the efficiency of a given impactor for different drop sizes as a function of only a few impaction parameters. The theory states the equations of motion for a droplet in the flow field around a single obstacle and subsequently introduces normalized dimensions in order to yield a single differential equation which principle depends on the flow field and the Stokes number /10,11/.

With the assumption of potential flow general solutions have been determined for basic obstacle types. From this, impaction efficiencies for spheres, rods, and jet impactors have been developed as a function of the Stokes number

$$St = \frac{\varrho_L \cdot d_T^2 \cdot V_0}{18 \cdot \eta \cdot D}$$

The Stokes number comprises the droplet properties droplet density ϱ_T and droplet diameter d_T, the kinematic viscosity of air η, and the reference dimensions reference velocity V_0 and reference length D (see figure 2a). Thus, the collector design affects the impaction process only through the parameters V_0 and D. In the case of the jet impactor, V_0 is the air velocity in the impactor's nozzle and D the diamter or the width of the nozzle (see figure 2b). The impaction efficiency $E(d_T)$ of a jet impactor for a given droplet diameter thereby becomes

$$E(d_T) = f(\ V_0,\ 1/D,\ 1/(s_D/D)\ )$$

As an additional parameter, the normalized distance between impactor plate and nozzle s_D/D affects the impaction.

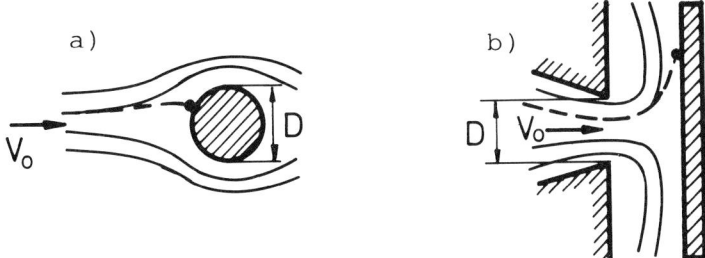

Figure 2. Impaction of a droplet a) on a cylinder and b) in a jet impactor (———streamline, – – – droplet trajectory)

An increase in the impaction efficiency is achieved by an increase of the jet velocity V_0, decrease of the nozzle width diameter D or a smaller distance s_D/D. Figure 3 illustrates this relationship.

In the case of the string or rod as impactor element, V_0 is the approach velocity of the air in an undisturbed region ahead of the obstacle and D the string or rod diameter. When strings or rods are arranged in a harp plane the normalized distance of the cylinder centers s_F/D enters as additional parameter. If batteries of more than one impactor planes in a row are set up, their number n also has to be considered /6/. The overall impaction efficiency of an impaction unit comprised from rods or strings is given by

$$E(d_T) = f(V_0, 1/D, 1/(s_F/D), n_H)$$

Thus, the impaction efficiency for a given drop size can be increased by a higher approach velocity V_0, smaller string or rod diamter D, narrower spacing of strings or rods, or increasing the number of impactor elemnts. Figure 4 shows the

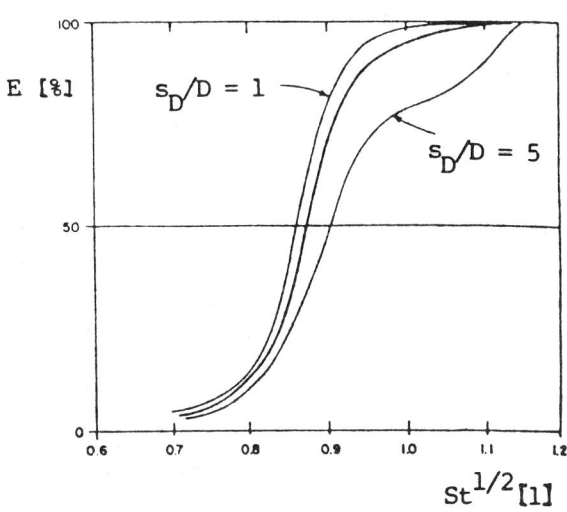

Figure 3. Impaction efficiency E of rectangular jet impactor as function of Stokes number St and relative distance s_D/D.
From: Mercer and Chow /12/

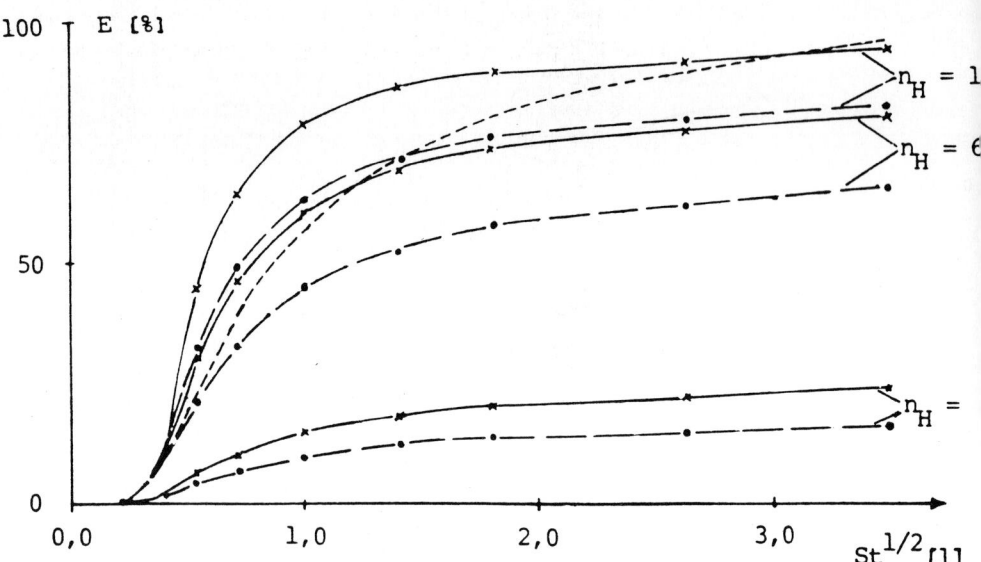

Figure 4. Impaction efficiency E of batteries with 1, 6, and 1 string impactor planes as function of the Stokes number St and the relative string distance s_F/D developed from string efficiency according to Langmuir and Blodgett /13/.
(- - - single string, — — $s_F/D = 6$, ———— $s_F/D = 10$)

dependencies of the overall impaction efficiency of an impaction battery developed from the single string efficiency. Practically, however, higher efficiencies will be achieved due to increased turbulence within the batteries /6/.

In the case of the cyclone a different impaction theory exists. For lack of space and the fact that cyclones have only been used occasionally in fog water collection, the interested reader should refer to the literature, e.g. /14/.

For the design of fog water collectors, the cut-off diameter of the sampled drop size spectrum is of prime concern (respresentative sampling). Unfortunately, two different definitions of the cut-off diameter exist in the literature. For jet impactors the size impacted with 50 % efficiency is referred to as cut-off diameter. For string and rod impactors the size collected with 0 % efficiency is called cutoff diameter. Here the latter definition is used.

Figure 5 illustrates the dependency of the cut-off diameter d_c from the velocity V_0 and the diameter D for a string of rod collector at air temperature $\vartheta = 20°C$. Similar plots can be derived for jet impactors.

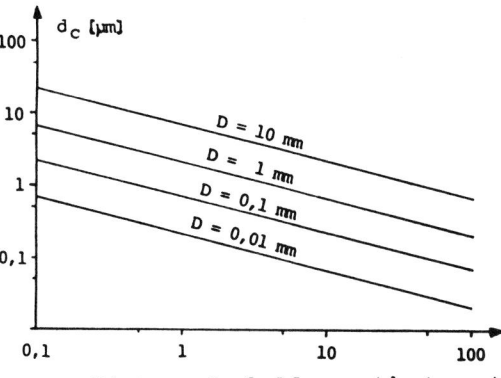

Figure 5. Cut-off diameter d_c for string or rod impactors as function of the approach velocity V_0 and the cylinder diameter D ($\vartheta = 20°$ C)

From figure 5 follows that cut-off diameters well below 0.1 μm can be achieved with the appropriate choice of D and V_0. However, limitations exist for both impaction parameters.
Two effects limit the choice of the air velocity V_0 around the impactor obstacle:
- resuspension of deposited fog water from the impactor: This may partially be counteracted by special impactor design.
- shift of the thermodynamic state of the air with its effect on the saturation condition in the impactor zone as a result of the pressure drop due to flow acceleration towards the obstacle and friction. Adiabatic expansion during acceleration would lead to considerable supersaturation at velocities V = 10 m/s. This is compensated by unknown frictional heating. Detailed research of that problem remains to be done yet. Clearly, moderate impaction velocities $V_0 \approx 10 \ldots 20$ m/s avoid significant changes of the saturation conditions in the impactor unit.

For the reference length D there exists a lower limit:
- Jet impactors with narrow nozzles of constant width are difficult to build and lead to large pressure drops. At a given flow rate the small width would have to be offset by extreme length of the nozzle to avoid near sonic velocities.
- Impactor strings that are exposed to mechanical stress require a minimum diameter. Chemically inert materials like teflon possess low mechanical strength and therefore are only available at string diameters D > 0.3 mm. Moreover, smaller string diameters would increase the residence time of fog water on the impactor strings because droplets on the string grow more slowly to a critical size necessary to run off to the base of the string.

The distance between impactor strings should be kept above 1 mm i.e. $s_F >$ D + 1 mm, to avoid that droplet bridges develop. These counteract fast droplet run-off.

The impaction theory proves to be an important tool for determination of sampling characteristics of fog water the collectors and to develop their design parameters. One has to bear in mind, however, that real impactor geometries differ from the ideal geometries of which the theory is based. Therefore, experimental calibration of fog water samplers should be conducted, especially in complex impactor designs.

The comparison of different collector types and individual designs reveals that no fog water sampler is clearly superior to others. Nor does anyone fulfill well all criteria for fog water collection /6/. The advantages of different sampler types are:
- defined air sampling rate, defined impaction conditions, and continuous sample water flow are attained by
  ⟶ vacuum collectors
- high air sampling rate and, consequently, high fog water sampling rate at moderate impaction conditions can be achieved by ⟶ vacuum collectors with string impactors
- high constant impaction efficiency down to submicron particles, sharp size cut, and small over-all dimensions of the collector can be realized with ⟶ jet impactors
- fast transfer of impacted fog water to the collection vessel can be obtained with ⟶ rotating collectors and
  ⟶ jet impactors
- omnidirectional sampling of fog air is achieved in
  ⟶ horizontally rotating collectors and
  ⟶ vacuum collectors with the sampling axis turned into the wind automatically or without preferential air sampling direction
- potential for automatisation possess all
  ⟶ vacuum collector types

Passive collectors cannot be recommended from the view-point of these criteria. No having power requirement and being cheap they may be applied in large numbers to investigate the spatial distribution of fog deposition.

Jet impactors as indicated above possess the best impaction characteristic of all collector types and also fulfill well some of the remaining criteria. However, when high fog water sampling rates are required for good time resolution or extensive chemical analysis replicate instruments have to be operated simultaneously to produce a sufficient sample water flow. Rotating collectors bear a security risk when impactors rotate at high velocities. They have to be attended by operation personnel during the entire sampling period. Lastly, cyclones work with high pressure drops. This leads to possible evaporation losses during impaction.

For the IVD fog program which is directed towards the potential of chemical input into forest stands via fog deposition, a high-volume fog water collector for chemical analysis was developed. The design of this vacuum collector with string impactors aimed at the improvement of existing samplers of that type in order to perform satisfactory with respect to all of the design criteria.

6. DESIGN OF THE IVD FOG WATER COLLECTOR

The principle design of the IVD high-volume fog water collector is plotted on scale in figure 6. Major elements are named and additional information of their dimensions or materials are given where relevant.

Furthermore, the path of air through the collector is indicated. At a rate of 840 m³/h fog air enters the collector through an circular inlet slot of adjustable height. It is formed by the cover held by three rods and the inlet nozzle which sits on top of the vertical tube. The cover prevents the intrusion of precipitation during collection and reduces contamination by wet and dry deposition between sampling periods. This is important when the device is operated automatically by a fog sensor. The inlet nozzle minimizes turbulence which would otherwise lead to significant deposition of droplets in that zone. It also helps develop a pipe flow profile. Due to the symmetrical arrangement of the air inlet with vertical axis, no

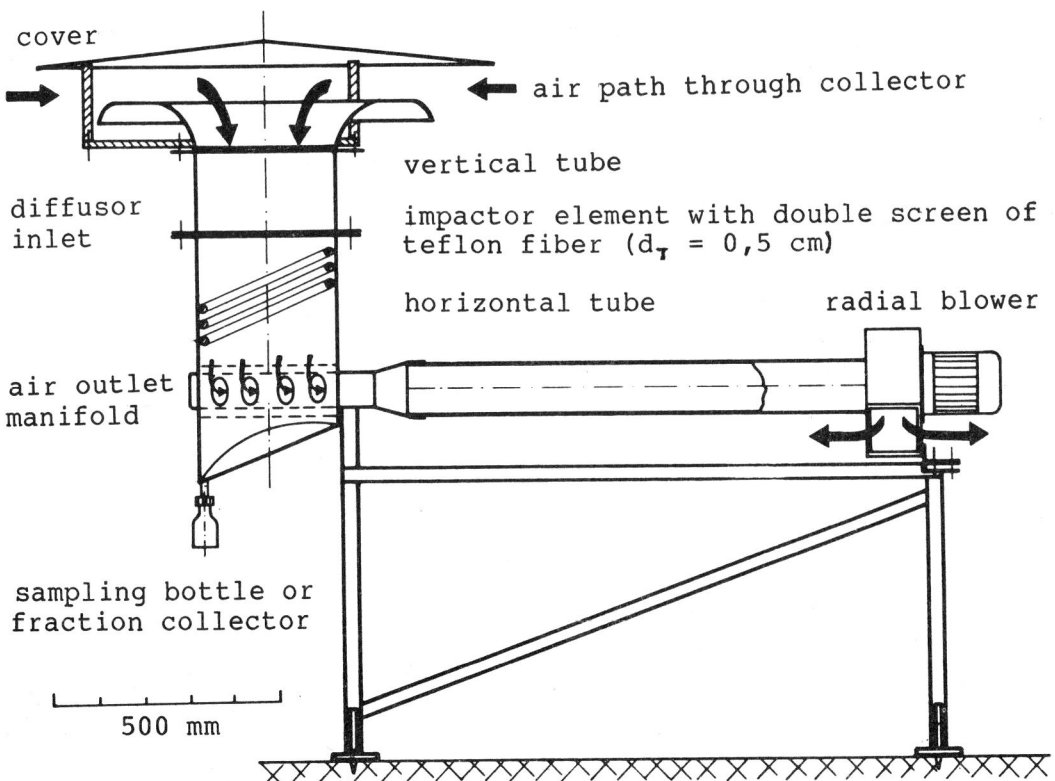

Figure 6. IVD high-volume fog water collector

preferential sampling direction exists. The air is drawn from shallow layer of air approximately 1.6 m above ground. Representative sampling of large fog droplets can be achieved in light and moderate winds, as an investigation of the inlet flow field showed /6/. In strong winds the air is likely to be diverted around the collector. Then large droplets that have entered the inlet will probably impact on the walls of the inlet nozzle or the vertical tube.

As the fog air progresses, it is deviated into a vertical axis and approaches the impaction unit which consists of a variable number of elliptical plastic rings with teflon fiber ($\emptyset$ 0.5 mm) strung around so that each impactor element builds two harp planes. The strings have a spacing of 2 mm in the harp plane. Each ring sits on 3 bolts through the vertical tubes wall and is inclined by 20°. There the separation of fog droplets from the gas phase takes place. After the passage through the impactor unit, the dried air leaves the vertical tube radially through eight holes distributed around the tube's circumference. These air streams are then reunited by the air outlet manifold designed with a radius increasing towards the horizontal tube so that air is drawn evenly through all openings. The dried air proceeds through the horizontal tube to the radial blower (400 W, 220 V). It provides the vacuum necessary for the air flow through the collector. Finally, the air is blown off several meters in a horizontal jet at about 0.8 m above ground. Thus short circuiting is almost avoided.

The path of impacted fog droplets to the sampling bottle starts at the teflon strings of the impactor rings. Due to coalescence, the droplets grow to a critical size for the run-off along the strings following gravitation and the drag force exerted by the air stream. From the base of the strings the sample water flows to the lowest point of the impactor ring and continues down the inner wall of the vertical tube reaching to the V-Shaped inclined bottom part welded into that tube. At it lowest point a standard 50 ml or 100 ml plastic bottle for sample water collection can be mounted. Alternately, the continuous sample water stream can be directed to a fraction collector or an on-line analytical unit.

Resuspension of coalesced drops from the impactor strings on their support rings does occur but only affects big drops. These, however, are unable to follow the air stream on its 90° diversion prior to its passage through the outlet openings and precipitate to the funnel shaped bottom part. In order to prevent losses of droplet running down the vertical tube wall 20 long tubes with the inside ends widened wre fit into the outle openings. They also concentrate the air stream to the center o the vertical tube. Thereby air and sample water flow are separated and possible interactions between gas and liquid minimized. Evaporation in that zone due to frictional heating can be ruled out because the pressure drop up to that point is negligible.

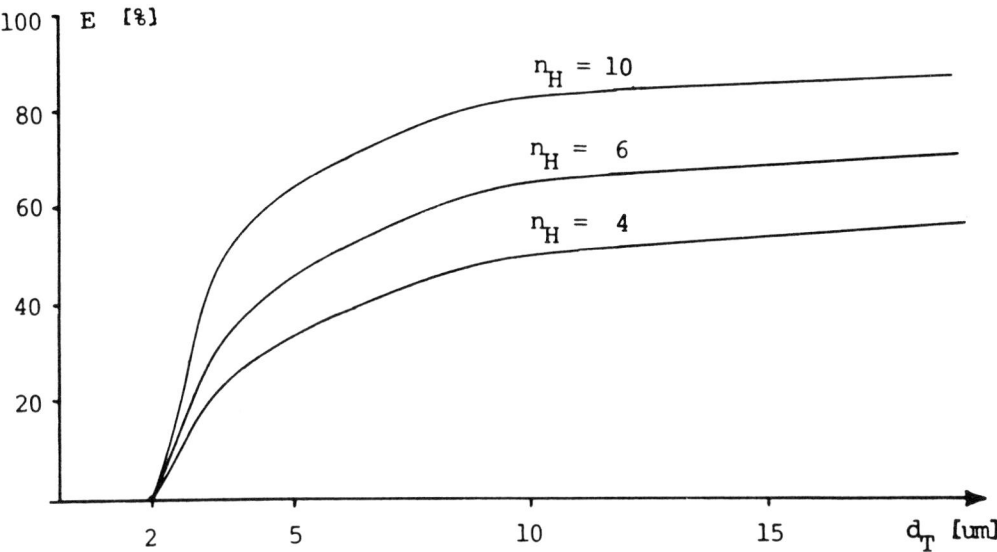

Figure 7. Over-all impaction efficiency E of the IVD fog water collector for 4, 6 and 10 harp planes as function of the droplet diameter d_T ($\vartheta = 20°C$, p = 1 bar)

Cleaning of all collector parts is done conveniently. For better accessibility, the vertical tube is divided into two parts; the impactor rings are inserted from the top and simply laid onto the support bolts. For effective cleaning of the impactor elements the use of an ultrasonic cleaning bath is recommended. The support structure of the collector is adjustable to uneven ground and easily put apart into several elements.

For this fog water sampler figure 7 shows the calculated impaction efficiency of the over-all configuration for 4, 6 and 10 harp planes corresponding to 2,3 and 5 impactor elements. An experimental calibration is planned for the future.

From figure 7 appears that complete droplet impaction cannot be attained with such a string impactor system. However, almost constant efficiency exists in the diameter range above $d_T \approx 10$ µm. Even droplets of $d_T = 5$ µm can be sampled with relatively high efficiency. Table II summarizes the design parameters. Additional features are:
- omnidirectional active sampling of fog air
- defined air sampling rate allowing a crude determination of the liquid water content
- continuous sample water flow with the potential for automatisation using a fog sensor and for on-line analysis of unstable chemical compounds.

TABLE II: Design parameters of the IVD fog water collector

Parameter	Value
Air sampling rate	$\dot{V}_L$ = 840 m^3/h
Approach velocity of droplets	V_o = 5 m/s
Crosssectional area of flow in impactor	A = 650 cm^2
Number of harp planes	n_H = 6
String diameter	D = 0.5 mm
Distance of centers of neighboring strings	s_F = 2.5 mm
Absolute cut-off diameter (E = 0%)	d_C = 2 m
Sample water flow rate	$\dot{V}_W$ = 50 ml/h

7. PRELIMINARY FIELD DATA

In late October 1984 the IVD fog water collector was operated on the site of the municipal waste water treatment plant of the City of Waldenbuch located in a mountainous area with extensive forerts south of Stuttgart. The collector was set up upstream from the plant thus facing the nocturnal winds down the valley. Fog samples were collected in two nights with radiation fogs. On Oct. 28 in intense fog of 80 m - 100 m visility at least 300 ml fog water were sampled between 1 a.m. and 7 a.m. Since the collection bottle was not emptied during the collection interval only the first and the last fraction of sample water could be analysed. The last fraction consisted of water backed up in the funnel part of the vertical tube. On Oct. 30 between 2 a.m. and 4 a.m. fog density increased steadily from ~ 300 m visibility. One sample was taken. Icing of the impactor string occured in the further course of the night. The results of the chemical analysis of these preliminary fog water samples are listed in table III and compared to concentration ranges of fo water constituents determined by Schrimpff et al /3/ using a passive collector and Stumm et al /15/ using a rotating arm collector.

These analyses show a rather high ion load of the fog water collected. Unusually high pH-values near neutral were found. The ion balance Σ (+)/Σ (-), i.e. the ratio of the sum of cation equivalents and the sum of anion equivalents, points to undetected cations, probably ammonium which was not analysed. Ammonia which is discharged in considerable amounts from anaerobic processes is known to be readily absorbed by acidic droplets thereby exerting an neutralizing effect. The high concentrations of magnesium and manganese could be the effect of contamination by aerosol emission from the treatment plant.

For this preliminary field test of the IVD fog water collector it can be concluded that works basically satisfactory. It is planned to develop a fog sensor to operate the sampler automatically and to obtain sequential samples with the help of a fraction collector. This could be further steps towards a continuous fog water analysis.

TABLE III: Chemical analysis of fog water samples from Aich valley near Waldenbuch compared to ranges given by Schrimpff et al /3/ and Stumm et al /15/

Quantity	10/28/84 first fraction	10/28/84 last fraction	10/30/84 before icing	Ranges in Schrimff et al /3/	Ranges in Stumm et al/15/
Sample volume in ml	100	35	40	–	–
Conductivity in µS/cm	234	199	277	69–1230	–
pH value	6.20	6.05	6.05	2.5– 4.2	2.3– 6.9
SO_4^{2-} in µeq/l	840	780	1140	20– 260	200–6000
NO_3^- in µeq/l	320	270	420	80–3000	100–6400
Cl^- in µeq/l	320	320	270	60–1500	150–5700
Na^+ in µeq/l	110	83	72	60– 970	–
K^+ in µeq/l	150	118	55	–	–
Ca^{2+} in µeq/l	300	180	265	50–2530	–
Mg in µg/l	780	–	–	–	–
Cd in µg/l	8	–	–	0.6– 36	–
Zn in µg/l	642	–	–	44– 118	–
Mn in µg/l	57	–	–	–	–
$\Sigma(+) / \Sigma(-)$	0.80	0.98	0.70	–	–

REFERENCES

/ 1/ Mayer, R. (1985): 'Verfahren zur Erfassung der Schadstoff-fuhr in Waldökosystemen' Staub **45** pp 267
/ 2/ Grunow, J. (1955): 'Der Niederschlag in Bergwald' Forstwissenschaftliches Centralblatt **74** pp 21
/ 3/ Schrimpff, E. et al (1984): 'Anwendung eines Grunow-Nebelfängers zur Bestimmung von Schadstoffgehalten in Nebelniederschlägen' Staub **44** pp 72
/ 4/ Whitby, K.T. (1978): 'The Physical Characteristics of Sulfur Aerosols. Atm. Environment **12** pp 135
/ 5/ Jiusto, J.E. (1974): 'Remarks on Visibility in Fog' J. Appl. Met. **13** pp 608
/ 6/ Dröscher, F.: 'Entwicklung eines Gerätes zum Sammeln von Nebelwasser' Diplomarbeit Nr. 2159, IVD, Univ. Stuttgart, 1984

/ 7 / Kunkel, B.A: 'Microphysical Properties of Fog at Otis Air Force Base' ATGL-TR-82-0026 Air Force Geophys. Lab., Hanscow, Mass., 1983
/ 8/ Jiusto, J.E., Carland Lala, G.: 'Radiation Fog Field Programs-Recent Studies' Atm. Sci. Res. Centr. - State Univ. New York (ARSC-SUNY) Publ. No. 869, Albany, N.Y., 1983
/ 9/ Nießner, R., and Klockow, D.: 'Atmosphärische Spurenstoffe in Niederschlagswasser: Konzentrationen und chemisches Verhalten Arbeitsmappe, VDI-Seminar: Chemische und physikalische Reaktionen von Spurenstoffen in der Atmosphäre, Stuttgart, 21./22.2.1984
/10/ Johnson, J.G.: 'Physical Meteorology' Technical Press, Cambridge, 1954, pp 223
/11/ Andre, K.: 'Bestimmung des Massenabsorptionsindex an Niederschlags- und Nebelwasserrückständen sowie an nicht aktivierten Aerosolpartikeln für den Spektralbereich 380-920 nm' Diplomarbeit Univ. Mainz, 1976
/12/ Mercer, T.T., and Chow, N.Y. (1960): 'Impaction from Rectangular Jets' J. Colloid Interface Sci. 27 pp 75
/13/ Langmuir, I., Blodgett, K.B.: Rep. No. RL 225, General Electric Res. Lab., Schenectady, N.Y., 1944/45
/14/ Weber, E., and Brocke, W.: 'Apparate und Verfahren der industriellen Gasreinigung' Bd. 1: Feststoffabscheidung. Oldenbourg, München, 1973, pp 46
/15/ Stumm, W. et al (1985): 'Der Nebel als Träger konzentrierter Schadstoffe' Neue Züricher Zeitung, Jan. 16, 1985 p 71

THE TEMPORAL DISTRIBUTION OF TRACE ELEMENT CONCENTRATIONS IN FOGWATER DURING INDIVIDUAL FOG EVENTS

Günther Schmitt
Institute of Meteorology and Geophysics
University of Frankfurt
Feldbergstr. 47
6000 Frankfurt/Main

ABSTRACT. Two sampling systems to collect fogdroplets are described. The results show the extremly high trace element concentrations in fogwater. Compared to rainwater in general the concentrations in fogwater show an increase by a factor of 10 - 15. A characteristical distribution of short time fluctuation of concentrations during single fog events is given. The measured concentrations strongly depend on the actual values of the liquid water content. Highest concentrations are found mostly at the beginning and the end of the fog period, when the liquid water content is reduced.

1. INTRODUCTION

Up to now the knowledge about acid fog and the chemical composition of the fogdroplets is based on only a few measurements, which are not representative. In addition the available results are not comparable due to different sampling methods. Nevertheless fog is one important pathway of removal of trace elements from the atmosphere. The mechanism of interception of fog droplets may be of great importance with respect to the atmospheric input into forest ecosystems. The interception of fog droplets can moisten canopy surfaces, thus increasing their effectivness in capturing trace gases and particles from the air. In forest areas with high fog frequency the total amounts of the intercepted fogwater can reach up to the yearly precipitation rates (Grunow 1955). For these reasons during the last few years fogwater chemistry is becoming an increasingly important part of atmospheric research. Main interests on our own investigations are focussed on the measurements of the temporal variations of concentrations in fogwater during individual events. Therefore an active sampling system was developed, which allows to collect several samples during individual fog episodes.

2. SAMPLING METHOD

Most fogwater sampling systems are based on the principle of inertial

impaction. By forcing the airstream to move around a barrier, fog droplets greater than a given radius will not be able to follow the streamlines. These droplets are impacted at the barrier. Collection efficiency and cut-off depend on the relative velocity between the airstream and the barrier as well as the dimensions of the barrier itself. With increasing relative velocity, the cut-off permits smaller droplets to be collected (Fuchs 1964).
In general the principal ways to collect fogwater can be distinguished into two main categories. The passive sampling methods use the natural relative velocity between the fog droplets and a stationary sampler (Black and Landsberg 1983, Mohnen 1980). In the opposite the active collection systems take the advantage of an increased relative velocity by rotating a collector through the fog or by sucking the fog droplets through the collector by means of a fan (Hoffmann et al. 1983, Katz 1980).

2.1 Passive method

Figure 1 shows a sketch of the developed passive sampling system. The collector consists of a teflon support structure and teflon strings of 0.3 mm in diameter, mounted every 3 mm around the periphery. In this case the teflon strings are used as the barrier in the airstream. The collector (S) is exposed in a height of 1.5 m above the ground, so that the air is enabled to pass the collector from all sides (Georgii and Schmitt 1985).
The fog droplets are impacted at the strings and grow up to larger drops. Finally they run down the strings and are collected in a 100 ml polyethelene sampling bottle. Using the above mentioned dimensions for the sampler and considering normal wind speeds, all droplets greater than 5 µm are impacted.
The amount of the collected fogwater naturally depends on the liquid water content (LWC) of the fog. In general an sampling time of 2 hours is necessary to collect a volume of 15 - 25 ml, sufficient for the chemical analysis. The whole sampling system is constructed in a way, that only during fog events the collector is exposed. For the remaining time the collector is kept in a metal cylinder as protection against contamination by rain and dry deposition. A fog sensor was constructed, comparing the actual temperature and the dewpoint (NS). As soon as the actual temperature reaches the dewpoint, adequate to an increase of the relative humidity up to 100 %, the fogsensor gives anelectronical signal to a motor (M) moving the collector out of its shelter (G). In case of rain, detected by an additionally sensor, the collector is moved back again into its shelter. A counter gives information about the sampling time.

2.2 Active method

Compared to the passive method the active system simulates higher relative velocities between fog droplets and collector. The sampling system has a powerful electromotor rotating an aluminium arm of 2 meter

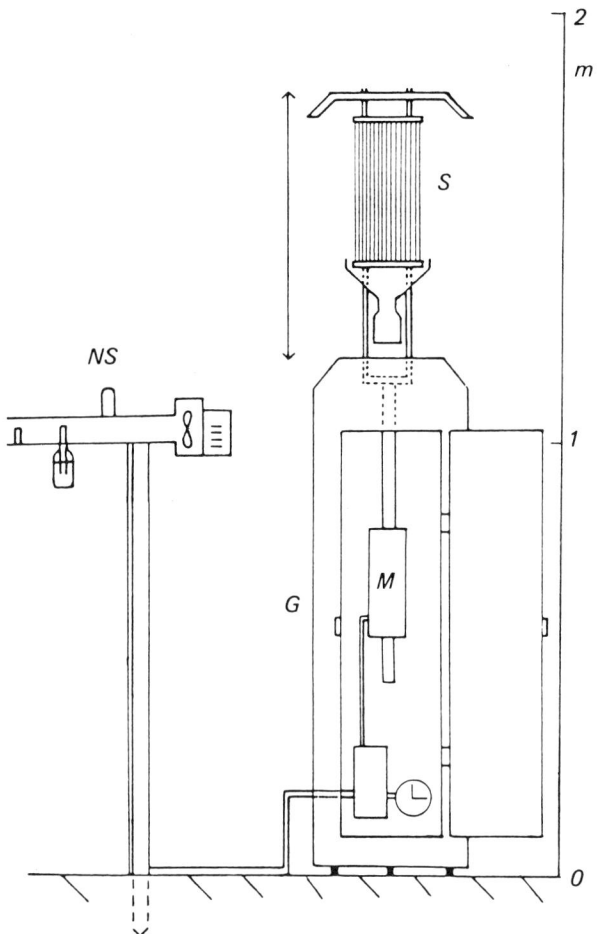

Fig. 1 Passive fogwater sampling system
A fog sensor, comparing the the actual temperatur and the dewpoint, gives an signal to a motor, which move the collector out of its shelter

length horizontally through the fog. At both ends of the arm collectors are installed (Fig. 2a). A sketch of the collector in detail is shown in figure 2b. Seven telfon bars, each 10 mm in diameter are mounted in three rows. Each bar has a sampling slot of 3 mm width (fig. 2c). During operation the collectors will turn into horizontal position due to centrifugal force. The fogdroplets are impacted in the sampling slots and are driven into the bottles. The main advantage is that previously impacted drops in the slots cannot be lost due to the high speed of the airstream. Also the evaporation of the sampled drops is reduced. The velocity of rotation can be adjusted electronically from 0 to 300 rpm, so that the velocity at the collectors reaches up to 30 m/sec. At this speed all droplets greater than 5 μm are impacted in the sampling slots.

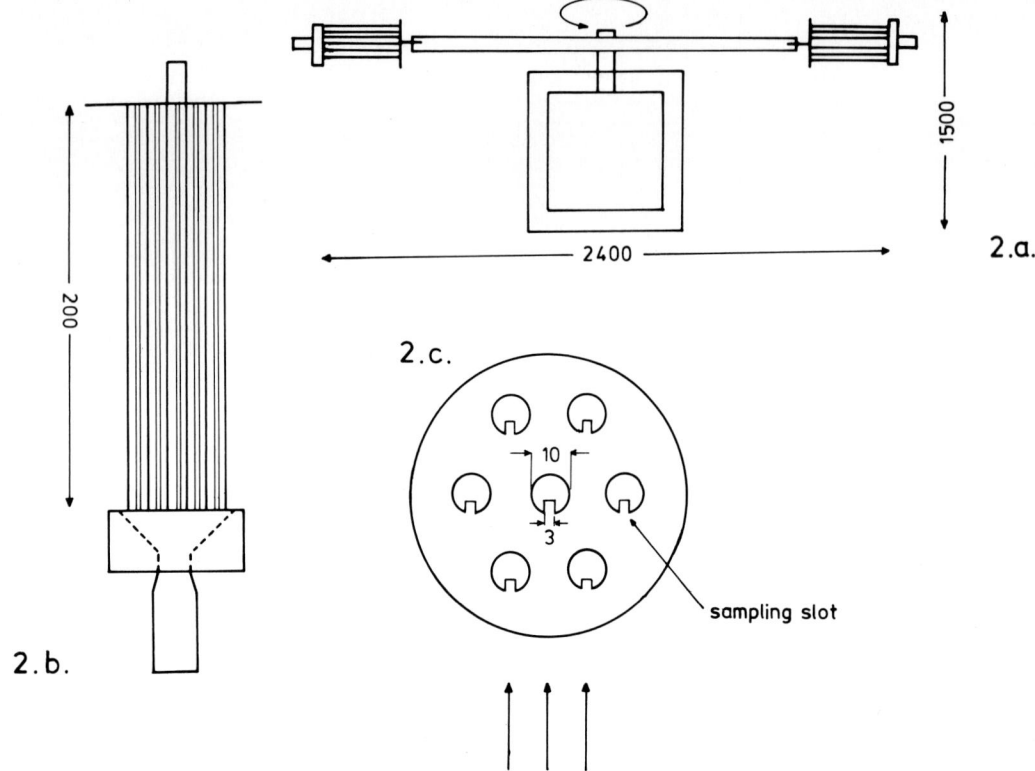

Fig. 2 Active fogwater sampling system
a.) general view   b.) collector in detail
c.) profile of the sampling slots

Using different relative velocities, the cut-off can be varied. Thereby it is possible to sample different droplet size fractions to obtain information about the chemical composition within these fractions.

Sampling times necessary to receive a sufficient volume of 20 ml fogwater range from 10 to 20 minutes, depending on the liquid water content of the fog. Using an active sampling system allows investigations on the temporal fluctuations of concentrations during individual fog events.

3. RESULTS

Both sampling systems are used sucessfully at a station on top of the mountain "Kleiner Feldberg" at an altitude of 800 m. The station is located 25 km northwest of Frankfurt. All fogwater samples are analyzed for pH, electrical conductivity, SO_4^- - S, NO_3^- - N, Cl^-, NH_4^+, Na, K, Ca, Mg and the heavy metals Pb, Fe and Cd.

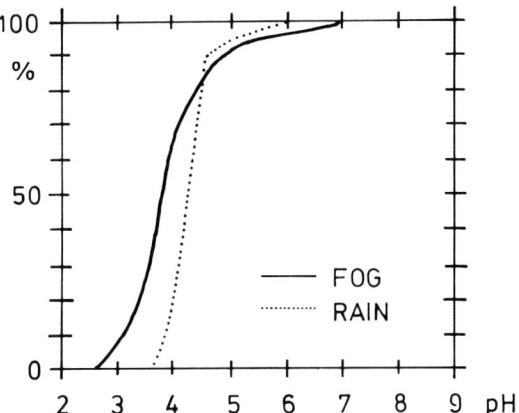

Fig. 3 Cumulative frequency distribution of the pH-values in fog and rainwater at the station "Kleiner Feldberg" (autumn 1983 - autumn 1985)

The pH-values are presented as a cumulative frequency distribution (Fig.: 3). The calculation includes 150 events. In comparison the pH-values in rainwater, sampled at the same station, are plotted in dashed lines. In fog 5 % of the measured values are below pH 3.0. The lowest pH found was 2.54, but also values between pH 5 to pH 7 were observed. In fog the 50 % value is pH 3.78, whereas in rainwater at the this station we find a mean value of pH 4.2. The cumulative distribution show, that in fog the pH are slightly lower than in rainwater. This is corresponding to higher element concentrations in the fog water. The distribution indicates the extremly wide variation of the pH-range, corresponding to a range in H^+ by more than four orders of magnitudes. In rain water the pH ranges only from pH 3.4 - pH 6.9.

Figure 4 shows the mean concentrations of the anions and ammonium for the period autumn 1983 - autumn 1985. The deltoides give the range of concentration. The peaks represent the maximum and minimum values, whereas the average values are indicated by the crossbeam. The highest concentration is found for the compound $SO_4^=$ - S, which ranges from 2.3 mg/l up to more than 80 mg/l. The averaged mean value is 13.2 mg/l. Again for comparison the concentrations measured in rain water are plotted as cross-bars. In rainwater the mean value is 1.37 mg/l. Compared to rain in fog there exist an increase by a factor of 10. Similar distributions are found for NO_3^- - N and Cl^-. For NO_3^- - N the mean value is 7.5 mg/l with highest concentrations up to 40 mg/l. For the compound Cl^- the measured concentrations range from 0.7 up to 30 mg/l with mean value of 6.5 mg/l. Also these compounds show an enrichmentfactor of 9 - 10 compared to the rainwater.
The result of the high concentrations along with the realtively high pH-values can be explained by the high contributions of the compound ammonium. The mean concentration is 8.6 mg/l. The maximum reaches up to

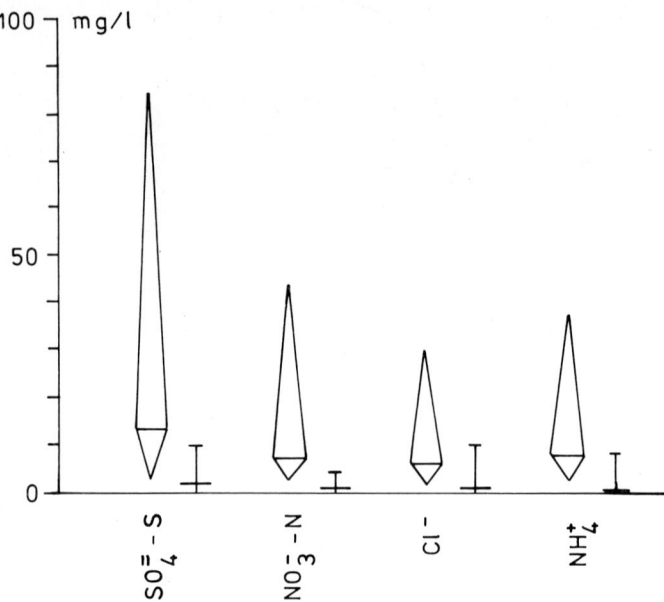

Fig. 4 Mean concentrations of the anions and NH_4^+
for the period autumn 1983 - autumn 1985
in fogwater and rain at "Kleiner Feldberg"
(deltoides: fogwater, cross-bars: rainwater)

40 mg/l. It seems, that different enrichment factors for the different compounds exist. While the anions in fogwater show an increase by a factor of 10, the concentration of ammonium is enhanced by a factor of 14. Up to now the elementspecific enrichmentfactors cannot be explained completly. One reason may be different anthropogenic sources or different physicochemical processes, such as certain reactions in the gaseous phase (Fuzzi 1986).

As mentioned above the active sampling system allows to collect different samples during individual fog events. As an impressive example not for very high concentrations, but for the typical temporal fluctuation of the concentrations during a single fog event the distribution of the anions, ammonium, lead, pH, electrical conductivity and liquid water content are shown in a case of a fog period collected at the seventh of october 1985. The samples are taken in half hours intervals. Sampling started when the visible range was less than 300 m. All element concentrations show a distinct equivalent distribution. At the beginning all measured compounds have high concentrations. In the following samples the concentrations decrease. For the following samples the concentrations increase and reach in some cases ($SO_4^=$ - S, Cl^-, Pb) the highest values of the whole fog event. The concentrations of the samples collected at 11.00, decrease again. Before the fog is disappearing, within the last samples, increased concentrations were observed. For the compound ammonium at the end of the fog episode the highest concentration was measured (14.45 mg/l), whereas in the first sample we found 12.3 mg/l.

It has to be pointed out, that for all investigated fog events we find high element concentrations in the beginning phase and at the end as a characteristic distribution for fog episodes.
For the analyzed heavy metals, as example the distribution of lead is given. The measured concentrations range from 120 µg/l to more than 300 µg/l.

While the same temporal distribution appears for the electrical conductivity (150 - 380 µS/cm), the pH-values show an inverse behavior. Samples with high element concentrations have low pH-values, decreasing pH-values are associated to increasing element concentrations. Nevertheless there is no explanation for the first pH values being relatively high in this case.

In addition figure 5 demonstrates the results of the samples taken by the passive sampling system at the same time indicated by the dashed lines. During the whole event only two samples could be collected. The sampling time was 3 hours for each sample. The measured element concentrations are in the same range as the samples collected by the active sampling system, but due to the longer sampling time the results produce only an integrated mean value of the temporal variation of the concentrations. It has to be mentioned, that for investigations on short time fluctuations of the concentrations within individual fog events passive collection systems are not satisfactory.

The observed variations of the concentrations and the pH-values depend on the actual amounts of the LWC. In general the LWC in fog ranges from 0.1 - 0.5 g/m³. Compared to clouds the LWC is much smaller (Juisto 1983). Also liquid water content is inversely proportional to the visible range in fog, although no absolutely relationship exists between the two parameters. The values of the LWC, reported in figure 4 were approximated by dividing the volume of the sampled fog water by the volume of air passed by the rotating collectors. At highest speed during one minute the two collectors pass a volume of 18 m³ air. The presentation shows, that low element concentrations and therefore high pH-values are only observed, when the LWC increases. In the opposite at low LWC the solution is concentrated leading to increasing concentrations.

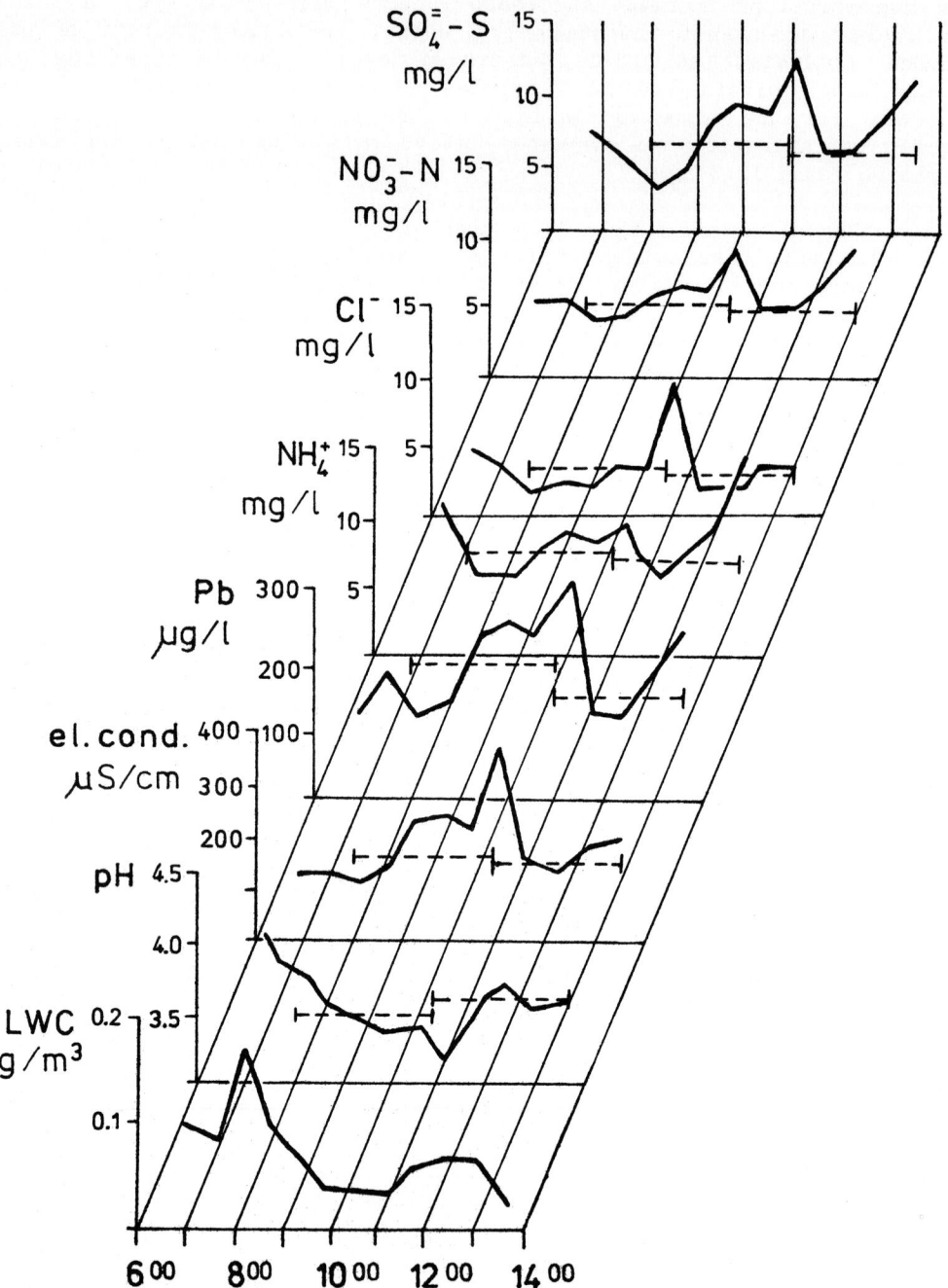

Fig. 5 Temporal distribution of the anions, ammonium, lead, pH, electrical conductivity and LWC during a single fog event (7.10.1985) at the station "Kleiner Feldberg" (solid lines: active system, dashed lines: passive system)

The dependence of the measured element concentrations in fog water on the liquid water content can be approximated by equation (1) (Junge 1963):

$$K = \frac{E * C}{L} \qquad (1)$$

K = concentration in fog water (mg/l)
C = concentration in air (mg/m³)
L = liquid water content (g/m³)
E = scavenging coefficient (0 < E < 1)
    the fraction of C, which enters the fog water

Generally high concentrations in fog compared to rain can be explained by equation (1) also. Near the ground, where fog mostly arises, the concentrations C of most of the important pollutants are enhanced compared to higher levels of the atmosphere, where clouds are formed.
In fog the scavenging coefficient E may be higher than in clouds, because in fog a larger fraction of small droplets can be found. Thereby the mean ratio of surface to volume of droplets is increased compared to clouds (Pruppacher and Klett 1978). Due to these differences, the incorporation and absorption of trace gases and aerosol particles is considered to be more effective in fog than in clouds. Finally in fog the concentrations will be incresaed due to the lower values of the LWC.

For further estimations we assume the concentration C of the trace subtances in the atmosphere and the scavenging coefficient E to be nearly constant during the time for the fog event. As result the concentration K in the fog water depends only on the actual values of the LWC. On the basis of the measured concentrations and the estimated LWC correlation coefficients were calculated.

In figure 6 the concentrations of the anions and H^+-concentration are plotted as a function of the LWC for a fog period in autumn 1984.
The absolute amounts of concentrations are much higher than in the case previously described before. The obviously strong dependence is given by correlation coefficients with high values of R > + 0.9. The estimations also indicate the extremly high concentrations, when the LWC is reduced. By decreasing the LWC for 50 % the concentrations will be increased by a factor of 2. For the investigastions on forest decline this effect has to be taken into account. Due to short time reduced LWC, as it is characteristic for the beginning and the end of a fog period, extremly high concentrations will occur. On the surface of the vegetation, when highly concentrated fog droplets were intercepted, the same effects may result in damage on the vegetation.

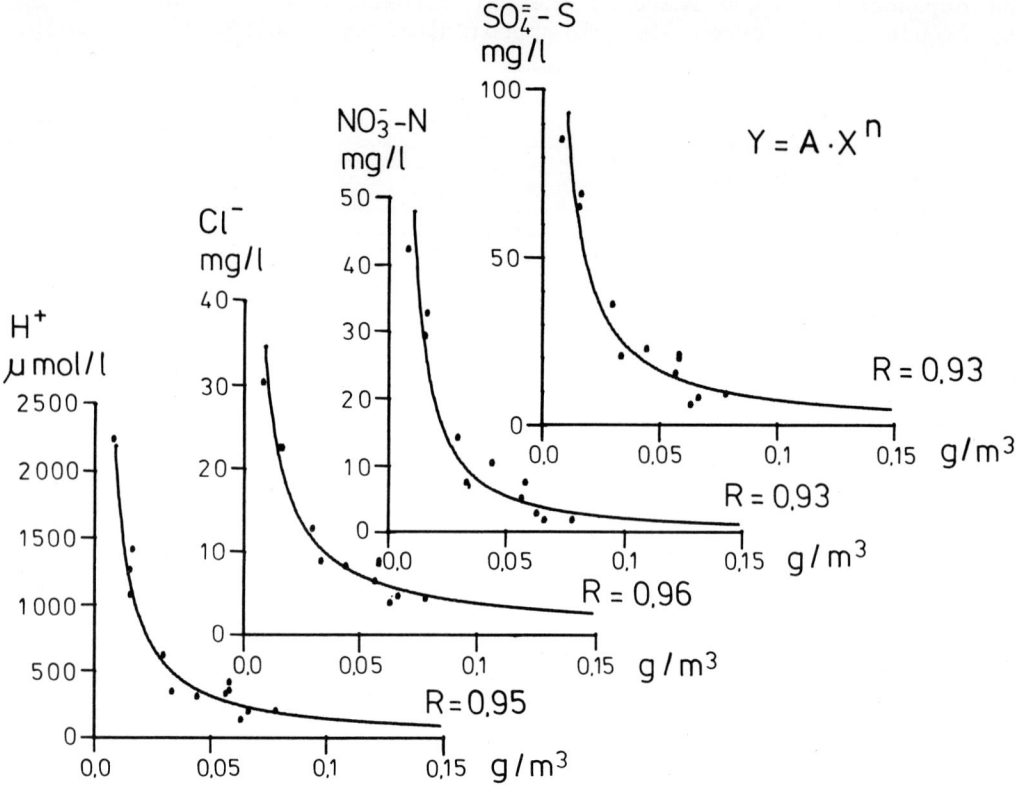

Fig. 6 Concentration of the anions and H⁺ as function of LWC during a single fog event (15.10.1984) at the station "Kleiner Feldberg"

By the data received from these measurements also a rough estimation on the abundant aerosol concentration can be produced. Compared to aerosol filter measurements the sample volume is defined by the total volume of air, passed by the collectors during rotation.

The product of the concentration K and the collected fogwater (Vw) is independent of the LWC. Dividing this product by the sample volume (Va) leads to concentrations of the elements (M) in the atmosphere:

$$M = \frac{K * Vw}{Va} \quad (2)$$

M = element concentration in the atmosphere (mg/m³)
K = concentration in fog water (mg/l)
Vw = volume of the sampled fogwater (l)
Va = volume of the filtered air (m³)

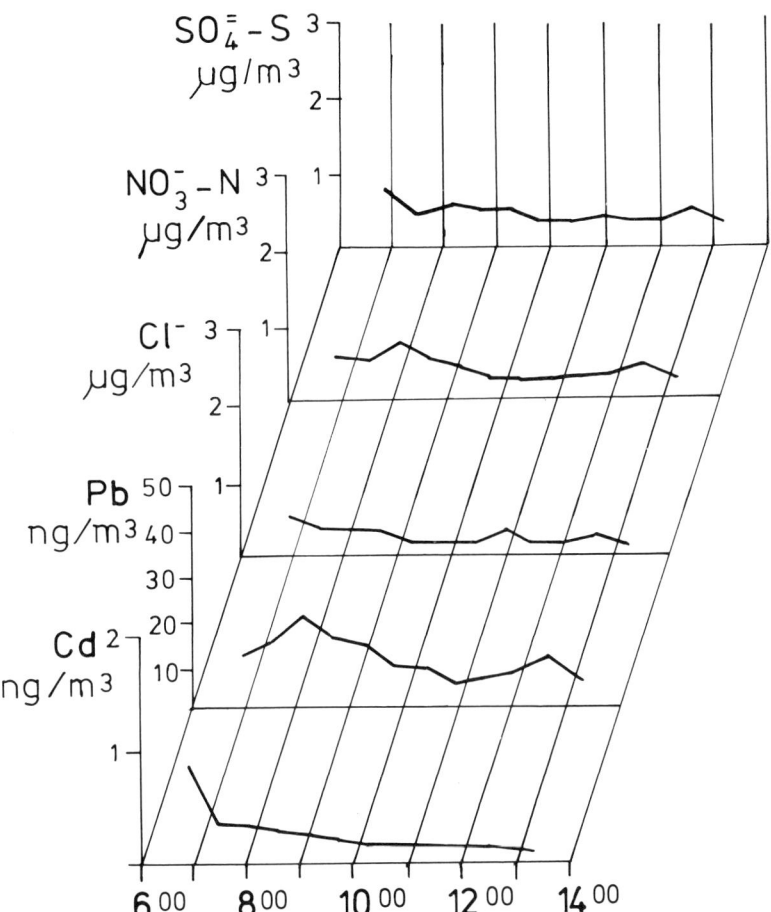

Fig. 7 Temporal distrbution of the element concentration M calculated by eq. (2) for a single fog event (7.10.1985) at the station "Kleiner Feldberg"

Figure 7 shows the calculations for the anions, lead and cadmium. The estimated concentrations of the anions range from 0.2 - 1.0 µg/m³. For the compounds lead and cadmium the concentrations varied in a small range from 10 - 20 ng/m³ respectively 0.5 - 1.0 ng/m³.

Compared to aerosol filter measurements at this station these results are in the same order of magnitudes (Georgii et al. 1982). The absolute amounts are always smaller, because the calculations by equation (2) only take into account element masses associated with the sampled fog droplets. Due to the independence on the LWC the concentration of M should be nearly constant during the fog event. Neverthelss the temporal distribution of M show a slight decrease during the fog event. This effect may be the result of removal of the elements from the atmosphere.

4. CONCLUSIONS

The measurements reported herein emphasize, that the investigations on fogwater chemistry are necessary for the complete understanding of the chemical processes in the atmosphere. Compared to rainwater the concentration in fog show generally an increase by a factor of 10 - 15. During individual fog episodes the variability of element concentration in fog is mainly determined by the actual amounts of the LWC. Extremly high concentrations were observed, when the LWC is reduced. Especially at the beginning and the end of a fog event the concentrations increase. A decrease of the LWC by half leads to an increase of the element concentrations by a factor of 2. The same effects may occur on plant surfaces, when fog droplets were intercepted by the vegetation. These results demonstrate the important role of fogwater chemistry with respect to investigations on forest decline.

A detailed research program on this field is planed in the future.

ACKNOWLEDGEMENTS

The investigations are sponsered by the Umweltbundesamt, Berlin under contract number 104 02 715.

REFERENCES

Black,H. and H.Landsberg (1983): A method for continuos records of the pH of low-level clouds. Final Rep Contr. DE-AS05 ER 6000044 College park, Maryland, 37 p.

Fuchs,N.A. (1964): The mechanics of aerosols, Pergamon Press, Oxford, p. 408

Fuzzi,S. (1986): 'Radiation fog chemistry and microphysics' Chemistry of multiphase atmospheric systems, ed.: W. Jaeschke, NATO ASI Series, G: Ecological Sciences, Vol 6 p. 213 - 226

Georgii,H.W. and G.Schmitt (1985):' Ergebnisse der Nebelwasseranalyse',Staub Reinh. der Luft, Nr. 6, p.260-264

Georgii,H.W., C.Perseke, E.Rohbock (1982): Feststellung der Deposition von sauren und langzeitwirksamen Luftverunreinigungen; Abschlußbericht Forschungsprojekt 104 02 600 im Auftrag des Umweltbundesamtes, Eigenverlag des Universtätsinstituts für Meteorologie und Geophysik, Frankfurt am Main, 1982, 205 p.

Gronow,J. (1955): 'Der Nebelniederschlag im Bergwald'. Forstw. Zbl., 74, p.21-36

Hoffmann,M. et al. (1983):' Design and calibration of a rotating
arm collector for ambient fog sampling'.
Precipitation Scavenging, Dry Depostion and
Resuspension, ed.: H.R.Pruppacher et al.
Elsevier Publ. New York Vol. 1, p. 125-137

Junge,C. (1963): Airchemistry and radioactivity
Academic Press, New York, 382 p.

Jiusto,J.E. (1983): Clouds, their formation,
optical properties and effects, Academic Press,
New York, p. 137 - 239

Katz,U. (1980):' A Droplet impactor to collect liquid water from
laboratory clouds for chemical analysis'.
Comm.a la VIII Conf. Intern. sur la Physique
des Nuages, Vol. 2

Mohnen, V. (1980): 'Cloud water collection from aircraft'
Atmospheric Technology 12, p.20-25

Pruppacher,H.R. and J.D.Klett (1980): Microphysic of clouds and
precipitation, Dordrecht, Reidel Publ., 714 p.

OBSERVATION ON FOG WATER COMPOSITION IN HAMBURG

P. Winkler
Deutscher Wetterdienst
Meteorologisches Observatorium Hamburg
Frahmredder 95
D-2000 Hamburg 65

ABSTRACT. Fog water was collected with an active sampler based on the impactor principle. A design has been developed which allows for a rapid separation of deposited water from the air stream. The fog water composition in Hamburg varies over several orders of magnitude: $SO_4^=$: 200-9000 ueq/l, NO_3^-: 200-7000 ueq/l, NH_4^+: 2-460 ueq/l, pH: 5.4-2.3, el. conduct.: 80-15000 uS/cm. This large variability is much higher as compared with precipitation and is caused by large variations in (1) trace substance availability, (2) liquid water content, and under certain meteorological conditions, (3) photochemical reactions producing HNO_3. These situations occur in relatively shallow fog into which radiation can penetrate so that photochemistry is induced.

1. INTRODUCTION

Whenever water condenses in the air either during cloud or during fog formation trace substances are incorporated into the droplets. In the beginning of the process water condenses on the aerosol particles. Once a droplet is formed it can take up further aerosol particles or absorb trace gases. The amount of absorbed gases depends on their solubility and subsequent reactions. Other important parameters which influence the concentration of material within fog droplets are the drop size distribution and the liquid water content (Fuzzi, 1983). If water condenses or evaporates very rapidly the dissolved material will become diluted or concentrated. In competition to this are the chemical reactions. These can change the chemical composition, whether they are important or not depends on mass transport (diffusion) and reactivity (Schwarz et al.,1981). Other possible changes on the trace substance content of fog water can originate from photochemical reactions (Chameides,1986). In the upper region of a fog or a cloud where radiation may penetrate, OH radicals can be produced which rapidly become dissolved in the droplets and induce further reactions.

In a city like Hamburg aerosol and trace gas concentrations can vary over wide ranges depending on wind direction, local weather situation and air mass. During fog sitiations with low wind speed and stable air local pollutants can accumulate and lead to high concentrations in fog droplets. At the Meteorological Observatory which is located at the northeastern border of the city either polluted or relatively clean air can reach the observation site.

2. FOG SAMPLER

Fog water is collected with a sampler operating on the impactor principle (see Fig. 1). The air is sucket at a rate of 95 m^3/hr through a twin nozzle behind which a specially designed deposition body is placed onto which the fog droplets are impacted. The deposition body has a vertically oriented hole in its center which is connected to the impaction surface by numerous small bored holes. The center hole and the instruments exit are connected by a tube so that a slight underpressure is applied sustaining a slight air flow through the capillary holes. The deposition surface has a small rim at each side preventing the deposited water from being riped off and carried away with the fast air stream. The fog droplets which are deposited coagulate and this water is sucket into the small holes due to capillary forces and under pressure and drains into the center hole from where its flows into the collection bottle. By this way a rapid separation of the collected water from the strong air stream behind the nozzle is reached and problems as evaporation or continuing reactions are minimized. Behind the nozzle

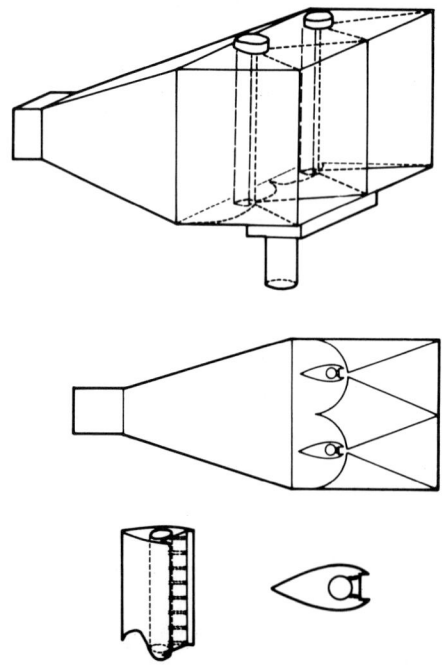

Figure 1. Schematic view of the fog sampler and the deposition body.

the air is guided by semicircular surfaces to the exit in order to avoid turbulence. The essential data are:

nozzle width 0.4 cm
nozzle length 2 x 12 cm
sucktion rate 95 m³/hr
50% cut off radius (calc.) 2.3 µm

The twin nozzle was chosen in order to keep the instrument handy. Up till now no experimental tests to study the efficiency and reliability of the instrument have been performed. At present (1985/86) the instrument is taking part at an intercomparison of fog samplers which is conducted at the University of Frankfurt.

3. OBSERVATIONS

Fog water has been collected since 1982 at the Meteorological Observatory. Sampling time was usually between 7 a.m. and the end of the fog. Although for this reason the material is not complete basic phenomena can be studied. At first results on pH-values and electrical conductivities are discussed using a pH-conductivity diagram (Fig. 2). Here the inclined lines represent constant fractions of free acid related to the total amount of dissolved material (Winkler, 1986). Given a droplet of pure diluted acid (e.g. HNO_3) of say pH = 3, a minimal conductivity of 360 µS/cm is adopted. The area above the 100% line is therefore forbidden. Addition of neutral salts to the droplet will increase only the conductivity while the pH remains constant. The line with number 10 represents a mixture of 10% pure acid and 90% neutral salts. The lines give information on the percentage of the cations available as H^+. If we have a given droplet with say 10% pure acid and condense more water so that a dilution by a factor 10 is caused the pH will be changed by one unit and the conductivity will be lowered by one order of magnitude. This means, that if the liquid water content of a fog changes due to condensation or evaporation without uptake or delivery of dissolved material the point in the diagram will move parallel to the inclined lines. If during a fog event acidity is formed due to uptake and oxidation of relevant gases we may expect a movement perpendicular to the inclined lines because the relative composition of the dissolved material was changed.

In fig.2 we recognize a large variation of pH values of different events over nearly 3 units. Accordingly, the electrical conductivity varies between 100 µS/cm and 15 µS/cm. This large variation seems to be a consequence of both, the variation in the amount of trace substances available for incorporation into fog droplets and the variation of the liquid water content. The variation is of similar magnitude as measured by other authors (Fuzzi, 1983; Jacob et al., 1985 and authors in Fig. 3). Therefore, no average or most frequent pH value can be derived. It can be stated, however, that on the average the acid fraction of fog is lower as compared with rain. In fog the contribution of H^+ ions related to total amount of cations is smaller than in rain or the dissolved material in fog is neutralized to a higher degree in spite of the fact that much lower pH values occur in fog than in the rain. In fig. 3 the average line for Hamburg

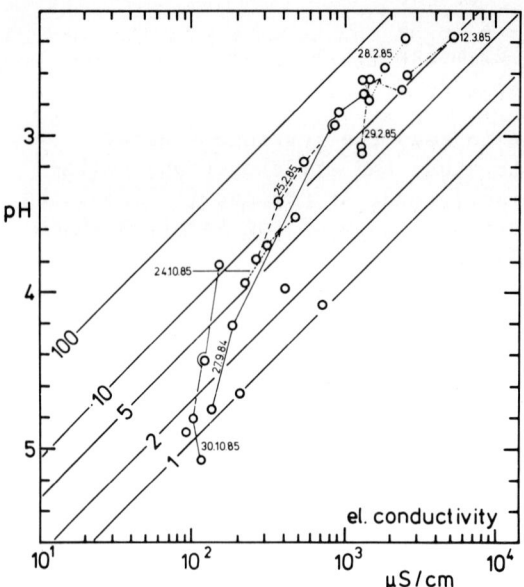

Figure 2. Acid fraction of fog water collected in Hamburg. Subsamples belonging to the same event are connected by this lines.

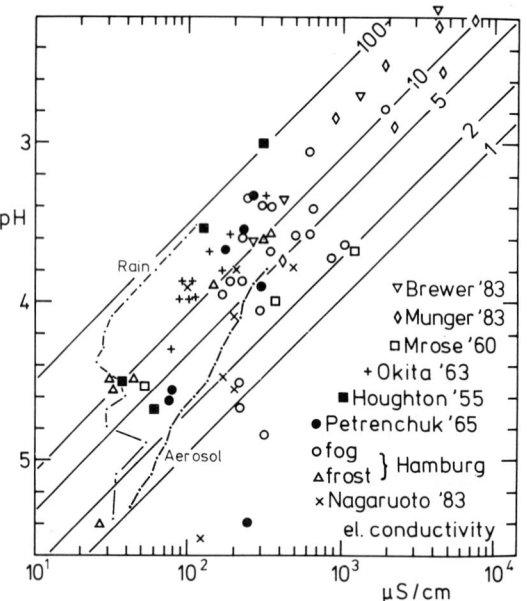

Figure 3. Acid fraction of fog water from Hamburg and other collection sites. The comparable average lines for rain water and aerosol particles are depicted.

precipitation is depicted. Its acid fraction is rather high (> 50%) at pH-values below 4.2. Such high acid fractions are occasionally observed in fog when the pH is below 3. The reason for the higher degree of neutralization in fog as compared with rain is twofold:
(1) the aerosol partciles acting as condensation nuclei usually show

acid fractions below 5 %. If water condenses onto such particles their acid fraction respectively relative composition remains constant as mentioned above. In fig. 3 the average acid fraction of aerosol particles is shown. For reasons of comparison these aerosol samples have been dissolved in such amounts of water (0.4 ml/aerosol mass from 1m^3) that a comparable dilution as in rain water was achieved. (2) The NH_3 availability is higher near the ground than in the level where raindrops develope, so that acidity which is formed from uptake and oxidation of gases is neutralized to a higher degree than the acid which is formed in rain drops in the cloud level.

During certain meteorological conditions a rapid formation of acid in fog water within short time periods was found. Up till now too cases have been observed on 27.9.84 and on 30.10.85. In both cases hourly samples have been collected. In fig.2 the points characterizing the subsamples which belong to one event are connected by thin lines. It can be seen that in both cases the first samples taken started with acid fractions around 1 % which rose to over 10 % two hours later. The highest increase in acidity occured within 1 hour in which the pH dropped by at least 1 pH unit. This cannot be explained by a decrease in the liquid water content because the points should have remained at the same line of constant percentage of acidity. Other fogs did not show such a development. During events at 24.10.85 or 12.3.85 the points shifted parallel to the lines of constant acid fraction which means that the pH variation can be understood in terms of changes in the liquid water content. One of the situations with rapid acidity

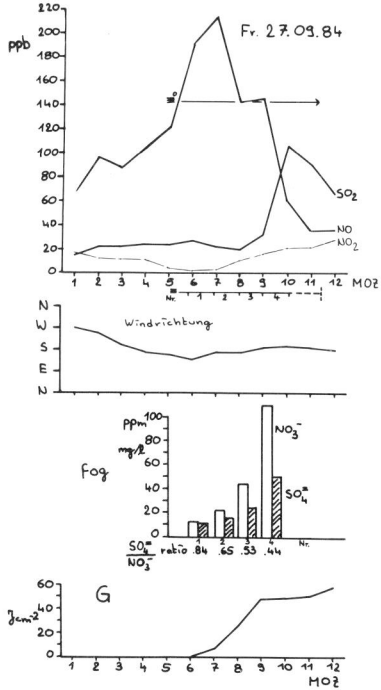

Figure 4. Detailed description of the development of the fog event on 27.9.84. For explanation see text.
1. Diagram: development of SO_2 NO and NO_2; ≡ indicates beginning of the fog event.
2. Diagram: wind direction.
3. Diagram: NO_3^- and $SO_4^=$ content of the four fog water samples.
4. Diagram: record of global radiation.
MOZ = mean local time

formation (27.9.84) is shown in greater detail in fig. 4. The upper part depicts hourly mean values of several trace gases. Four fog samples have been collected between 6 and 10 a.m. as indicated below the abscissa. The wind direction remained between 180 and 270 degrees, a sector from which a maximal influence of local pollutants from the city of Hamburg can be expected.

The third diagram shows the concentrations of $SO_4^=$ and NO_3^- analyzed by ion chromatography, and beneath the abscissa the ratio $SO_4^=/NO_3^-$. In comparison with fig.2 it can be seen that the rapid decrease in pH between sample 2 and 3 is caused by a rapid increase of $SO_4^=$ and NO_3^-, the NO_3^- increase being more pronounced than the $SO_4^=$ increase. As a consequence the $SO_4^=/NO_3^-$ ratio drops from 0.84 at the beginning to 0.44 at the end of the event. The pH decrease is therefore mainly caused by HNO_3 formation and uptake into the droplets. In the lowest diagram the raise of the global radiation is shown indicating that the fog was relatively shallow so that radiation could penetrate easily. On 30.10.85 the development was similar with a rapid pH decrease between sample 2 and 3. again caused by HNO_3 formation in a shallow fog. In the other cases (28.2.85, 12.3.85, 24.10.85) the HNO_3 formation not occured. The fogs had a greater vertical depth so that radiation could not penetrate to the ground. In these cases a lifted fog remain after fog dispersal at the surface. Therefore it can be supposed that HNO_3 is formed in the interstitial air due to photochemical reactions and subsequently incorporated into the fog droplets (Chameides, 1986; Platt et al., 1984). Because radiation cannot penetrate deeply into clouds or fog this process can only be observed during occasions with shallow fog. Nevertheless it is likely to occur in the upper region of a fog layer or in the upper region of clouds.

4. DISCUSSION

The concentration of the various constituents in fog water varies over several orders of magnitude. At Hamburg the following values have been found:

	pH	el.cond. /uS/cm	$SO_4^=$	NO_3^- /ueq/l	NH_4^+
fog	2.3-5.4	100-15000	200-9000	200-7000	2-460

These are at least 10, times higher as compared with rain water. Because of the high variability of concentrations in fog it is not possible to calculate average values. In order to do so, the measured concentrations should be recalculated for a certain liquid water content (e.g. 0.1 g/m^3). Unfortunately, during our studies determinations of the liquid water in fog has been started only recently, so that not enough material is available to present data in such a normalized manner. It seems necessary for future work, however, to present data normalized to a certain liquid water content in order to be able to compare the data from different locations or to be able to calculate average values for one site. As a recommendation data on fogwater com-

position should be presented in the following form:
(1) measured original concentrations
(2) normalized concentrations (e.g. for liquid water content = 0.1 g/m^3) together with the range of observed liquid water content
(3) averages of the normalized concentrations together with standard deviations and range of observed liquid water content.

The determination of the liquid water content is not easy because standardized measuring procedures are not available. Active fog samplers as the here presented impactor sampler allow the determination of the liquid water content as long as the cut off radius is low enough and the cut off function is steep. Passive fog samplers should be operated together with an instrument which determines the liquid water content. By employing such a procedure the scatter in data caused by variations in the liquid water content can be eliminated so that only the variations originating from different air chemical conditions will remain.

If no information on liquid water content is available the use of the pH- el. conductivity diagram is helpful because pH values for a standard el. conductivity can be easily derived. If one compares thus derived fog-pH-values with precipitation-pH-values one recognizes that the normalized pH of fog usually is higher than that of the rain.

As we have seen from figs. 2 and 3 the concentration of substances in fog water can vary over 3 orders of magnitude. If we assume, that about one order of magnitude is on account of the variability of the liquid water content a difference of two orders of magnitude remains on account of variations in air chemical conditions. It can be concluded from this rough estimation that the chemical composition of fog water respectively water of clouds in contact with mountains depends much stronger upon the local situation than it is the case with precipitation. Whether or not a plume or other strong sources of trace gases are present, aerosol content is high or low, fog is persistent or not, and so on will cause a "dirty" or a "clean" fog. A fog forming on a meadow in stable air without wind can be very clean. Downwind of a single stack in a valley where the plume is in contact with cloud or fog, the droplets can be very polluted while some hundred meters aside the droplets can be rather clean. Such circumstances may be of great importance for effects of cloud- or fogwater on plants or soil.

5. CONCLUSIONS

(1) In fog water chemical species can become highly concentrated, 10 - 100 times more than in precipitation. The composition of fog water is much more inhomogenious as compared with precipitation.
(2) The chemical composition should be presented normalized to a certain liquid water content (e.g. 0.1 g/m^3) in order to reduce the variability in the composition to the air chemical circumstances and to eliminate the variations of substance concentration which originate from different liquid water contents. This allows for a better comparability of the data from different places.

(3) In the case of acidity a pH-el.conductivity-diagram is helpful in this context. Shifting the pH values along the lines of constant acidity so that a conductivity as measured in precipitation is reached the resulting pH is higher in fog than in precipitation. This means that the dissolved material in fog is usually neutralized to a higher degree than it is the case for rain.

(4) Besides the air chemical situation and the liquid water content, radiation may have a marked influence on acidity formation. Under certain meteorological conditions radiation may stimulate HNO_3 production followed by rapid uptake into fog droplets. A lowering of the pH value by more than 1 unit within one hour has been observed.

6. LITERATURE

Brewer, R.L., Gordon,R.J., Shepard, L.S., Ellis, E.C.: 'Chemistry of mist and fog from the Los Angeles urban area'. Atmosph.Environment 17 (1983) 2267-2270.

Chameides, W.L.: 'Photochemistry of the atmospheric aqueous phase'. In: Chemistry of Multiphase Atmospheric Systems, ed. by W.Jaeschke, Nato ASI Series G6, Springer-Verlag Berlin, Heidelberg (1986) 369-414.

Fuzzi, S., Orsi, G., Mariotti, M.: 'Radiation fog liquid water acidity at a field station in the Po valley'. J. Aerosol Sci. 14 (1983) 135-139.

Houghton, H.G.: 'On the chemical composition of fog and cloud water'. J. Meteorology 12 (1955) 355-357.

Jacob, D.J., Waldman, J.M., Munger, J.W., Hoffmann, M.R.: 'Chemical Composition of Fogwater Collected along the California Coast'. Environm. Sci. Technol. 19 (1985) 730-736.

Mrose, H.: Measurement of pH and chemical analyses of rain, snow and fog-water'. Tellus 18 (1966) 266-270.

Munger, J.W., Jacob, D.J., Waldman, J.M., Hoffmann, M.R.: 'Fogwater chemistry in an urban atmosphere'. J. of Geophys. Res. 88 (1983) 5109-5121.

Nagamoto, C.T., Parungo, F, Reinking, R., Pueschel, R., Gerich, T.: 'Acid clouds and precipitation in eastern Colorado'.Atm. Environm. 17 (1983) 1073-1082..

Okita, T.: 'Concentration of sulfate and other inorganic materials in fog and cloud water and in aerosol'. J.Met.Soc.Japan 46 (1968) 120-127.

Petrenchuk, O.P., Drozdova, V.M.: 'On the chemical composition of cloud water'. Tellus 18 (1966) 280-286.

Platt, U.F., Winer, A.M., Biermann, H.W., Atkinson, R., Pitts,Jr.,.J.N.: 'Measurement of Nitrate Radical Concentrations in Continental Air'. Environ. Sci. Technol. 18 (1984) 365-369.

Schwartz, S.E.,Freiberg, J.E.: 'Mass transport limitations to rate of reaction of gases in liquid droplets: Application to oxidation of SO_2 in aqueous solution'. Atmosph. Environment 15 (1981) 1129-1144.

Winkler, P.: 'Relations between aerosol acidity and ion balance'. In: Chemistry of Multiphase Atmospheric Systems, ed.by W. Jaeschke, Nato ASI Series G6, Springer-Verlag Berlin, Heidelberg (1986) 269-298.

Case-Studies

TEMPORAL VARIATIONS OF TRACE SUBSTANCES DURING INDIVIDUAL RAIN EVENTS

Renate Zimmermann
Institute of Meteorology and Geophysics
University of Frankfurt/M
Feldbergstr. 47
6000 Frankfurt am Main
Federal Republic of Germany

ABSTRACT. Rainwater was collected simultaneously at 3 sites by automatically operated volume-based sequential samplers. Samples were analysed for conductivity, pH, $SO_4^=$, NO_3^-, Cl^- and soluble and insoluble fractions of Pb, Mn, and Fe. Rain from frontal precipitation system shows a regular pattern with high concentration of pollutants at the onset of rain, declining subsequently to constant, low concentration. During rainshowers the concentration of pollutants in rainwater is highly variable, reflecting the temporal variations and inhomogeneities of major microphysical and dynamical parameters. Washout was found to contribute more significantly to total wet deposition at the urban sampling site than at the more remote sites. Furthermore, the fraction of pollutants removed by washout was found to be larger for the heavy metals than for the anions sulphate and nitrate.

1. INTRODUCTION

Rain samples, collected on a sequential basis contain some information on the origin and mechanism of incorporation of their trace constituents. Sequential samples collected during different characteristic situations (e.g. convective showers vs. precipitation from stratified clouds) may be used to study the influence of meterological and microphysical parameters in the scavenging processes.
De Pena (1984) has found an inverse relation between the intensity of rain and the concentration of trace constituents in the rain samples collected, wheras the measurements of Kins (1980) and Martin and Barber (1984) could not show a clear realation. In warm front precipitation the concentration of trace constituents in rainwater generally is found to be highest during the initial phase of precipitation, declining rapidly until a level is approached which remains rather constant during the course of the rainfall (Kins 1980). The initial peak is interpreted mainly by these two mechanisms: a.) below cloud scavenging (washout) of pollutants, which are present at highest concentrations during the onset of precipitation and b.) increased evaporation of droplets below the

cloud base due to large undersaturation at the beginning of the rainfall. From several rain events sampled from stratified clouds the fraction of pollutants removed by in cloud scavenging (rainout) is estimated here. This estimate is based on the assumption that the concentration level approached after some time is due to rainout only, since all the material which could be subject to washout is removed by that time. The fraction removed by washout is estimated from the remainder by considering the contribution of below cloud evaporation.

Sequential rainwater samples have been collected simultaneously at 3 sites, differing from each other in altitude and pollution burden:
a.) in the city of Frankfurt (110m alt., residential and commercial area), b.) Königstein (rural, 520m alt.) and c.) Mt. kleiner Feldberg (rural, 840m alt.). A detailed description of the sites can be found by Georgii et al.(1982).

2. EXPERIMENTAL

Sequential sampling of a rainfall event can be carried out by spacing the samples on either a time (Dawson 1978, Raynor and Hayes 1983) or a sample volume basis (Coscio et al. 1982, Seymour and Stout 1983, Asman 1980). We have separated the rainwater during collection automatically into samples of 20ml volume, since this gave a good separation both at low and high intensities of precipitation and was sufficient material for the analysis. Furthermore a bulk sample was collected during each precipitation event. A detailed description of the sequential sampler was given by Kins (1980). Electrical conductivity and pH of the samples were measured as soon as possible after collection. Soluble and insoluble compounds present in the sample were separated by filtration, using membrane filters. In the filtrate, $SO_4^=$, NO_3^- and Cl^- were analyzed by Ion-Chromatography. The metals Pb, Mn and Fe were analysed by Atomic Absorption Spectrometry, both in the filtrate and in the residue of the filtration.

3. RESULTS AND DISCUSSION

3.1. Average concentrations in bulk samples

Fig. 1 shows a comparison of the average concentrations of elements and ions in the bulk samples collected simultaneously at Frankfurt, Königstein and Mt. kleiner Feldberg. The data for Frankfurt and Königstein represent averages of 12 individual precipitation events, at Mt. kleiner Feldberg 6 events were sampled. The samples were collected from February 1984 to April 1985. The concentrations of metals given in Fig. 1 include both the soluble and insoluble fraction. For comparison among the different sites and compounds, the data for each site are given as percentage-fraction of the value at Frankfurt.

For all compounds a decline is observed from the urban site at Frankfurt towards the rural sites at Königstein and Mt. kleiner Feldberg. This decline is most pronounced for the heavy metals, particularly for Pb. The difference between Königstein and Mt. kleiner Feldberg (8 km distance) generally is somewhat less pronounced than between Frankfurt and Königstein. For the anions in rainwater, the variation between urban and remote site is smaller, indicating a more homogeneous distribution of their gaseous and particle-bound precursors in the atmosphere.

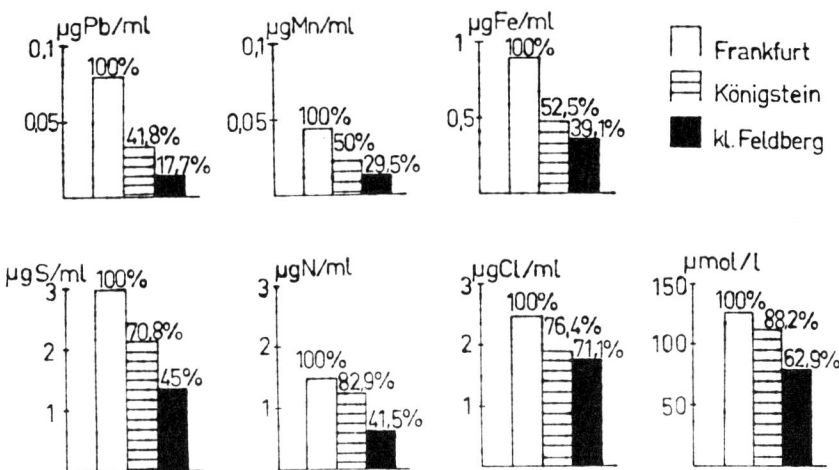

Fig. 1: Average concentrations of elements and ions at Frankfurt Königstein and Mt. kl. Feldberg

3.2. Variation of the concentration during individual events of precipitation

Due to their genetic mechanism and microphysical properties, rainclouds can be categorized into those associated with front systems and those originating from convective systems. These categories can also be distinguished from the temporal characteristics of the trace constituents found in the precipitation.

Precipitation from front systems
In Fig 2-7 the concentration of Pb, $SO_4^=$ and protons in sequential samples collected simultaneously at Frankfurt and Königstein during 5 individual frontal rains is plotted versus the rainfall amount.
Precipitation sampled in Frankfurt:
For Pb (Fig. 2), the highest concentration is found during the initial phase of precipitation, ranging from 0.08-0.3 µgPb/ml. Within the first 1- 2mm (2-5 samples) of rain the concentration rapidly decreased to 0.02-0.06 µgPb/ml, remaining at this level with only minor variations. Secondary maxima, like on 28/5/85 are caused by interruptions of the rainfall, subsequent accumulation of atmospheric pollutants and removal by washout when the rainfall starts again. The curves for Mn and Fe are similar to the one of Fig. 2 for Pb. The peak concentrations for Mn and Fe range between 0.03-0.1 µgMn/ml and 0.5-1.6 µgFe/ml, respectively.

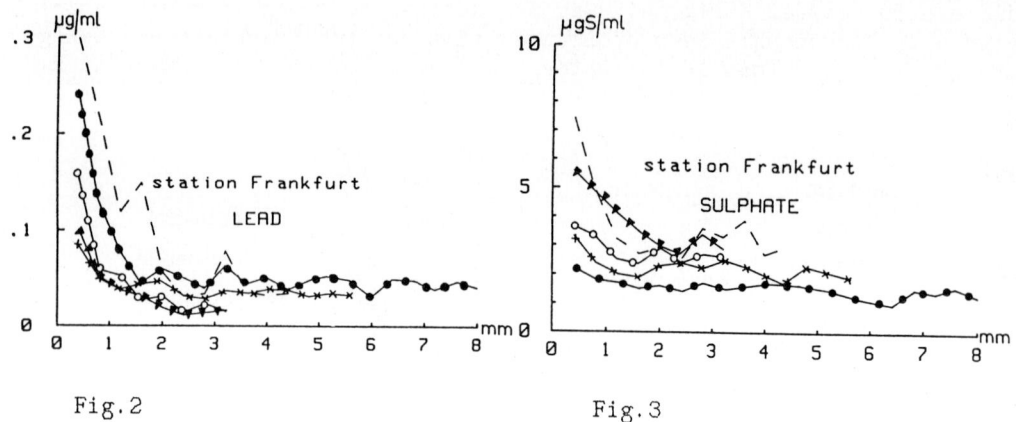

Fig. 2   Fig. 3

Sulphate, displayed as a representative of the anions in Fig. 3, shows less pronounced initial peaks. Characteristic peak concentrations of the anions ranged between 2.1-7.4 µgS/ml for sulphate, 1.2-4.1 µgN/ml for nitrate and 1.8-4.6 µgCl/ml. The protons in rainwater (Fig. 4) show a markedly different behaviour compared to the anions. The peak of protons is not found in the initial phase, but later during the course of rainfall. It is suggested, that neutralization of acidic compounds by soil dust-derived coarse alkaline particles which are removed from the atmosphere during the initial phase of rain, is the reason (Seymour and Stout 1983). The concentration of H^+-ions in rainwater is highly variable, fluctuating by as much as 120 µmol/l during a particular rainfall event.

The electrical conductivity of the rainwater samples was also found to be highly variable. As for many other parameters, the peak of conductivity was recorded at the beginning of rainfall. After the initial peak is passed, the major fluctuations in conductivity parallel those of H^+-ion concentration. The conductivity recorded during five frontal rains ranged between 10 and 105 µS/cm.

Precipitation sampled in Königstein:

As a representative of the heavy metals analyzed, Fig. 5 gives the concentration of Pb in sequential samples from Königstein, versus the rainfall amount.

Generally, both the peak concentration and the variability during the subsequent phase of decline were found to be distinctly smaller than for Pb at Frankfurt. The variability after 1-2mm of rain frequently was smaller than the analytical precision. The peak concentration of the heavy metals ranged between 0,05-0,16 µgPb/ml, 0,012-0,063 µgMn/ml and 0,3-0,9 µgFe/ml. Toward the end of the precipitation event, the concentration of heavy metals at Königstein approached the final values found at the site in Frankfurt.

For the anions, Fig. 6 shows $SO_4^=$, the peaks at Königstein were lower than at Frankfurt too, but the difference was less pronounced than for the heavy metals. For the anions, the overall variability and final concentrations were similar among both sampling sites.

The concentrations of H^+-ions (Fig. 7) ranged between 20 and 215 µmol/l. The concentration of protons was distinctly higher at Königstein than at Frankfurt, except during the event on 7/4/85. This may be due to a lower concentration of alkaline soil-derived mineral dust in the forested region at Königstein. The electrical conductivity in the samples from Königstein was lower than at Frankfurt, varying between 20 and 80 µS.

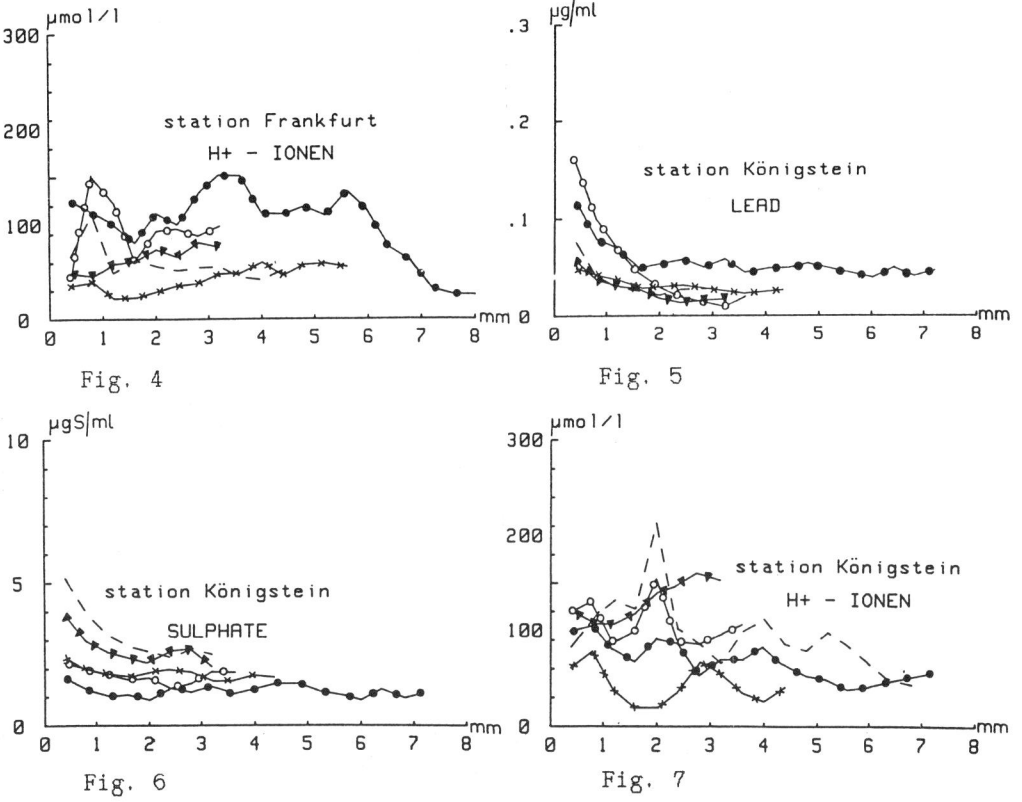

Fig. 2-7: concentration variations of Pb, $SO_4^=$, H^+-ions during five frontal rains
- ▼—▼ 20.5.84
- ○—○ 9.10.84
- — — 28.5.84
- ●—● 7.4.85
- ✶—✶ 14.9.84

Precipitation from convective clouds

Five convective showers were sampled simultaneously both at Königstein and Frankfurt and were subsequently analyzed for the parameters given above. Fig. 8-14 show the concentrations of Pb, $SO_4^=$, NO_3^- and H^+-ions in the sequential samples at both sites.

In frontal rains, the sequential samples from convective rains do not exhibit the regular pattern found for anions and heavy metals in frontal rains, with an initial peak and subsequently declining concentrations. For all compounds analyzed the variability both within a specific event and among different precipitation events was found to be very large. At both sites the highest concentrations were usually recorded during the course of the showers or towards their end. For most of the showers

sampled, a general trend with the concentrations decreasing as the amount of precipitation increases was not observed.
The concentrations of sulphate and nitrate mostly parallel each other. Maxima of the concentration often correlate with minima of the precipitation intensity.
For the protons, the minimum-concentration generally was recorded in the first sample collected. As stated above, neutralization by alkaline particles may be the reason.
The concentrations of heavy metals and anions were not markedly different among both sites, unlike it was observed during frontal rains.

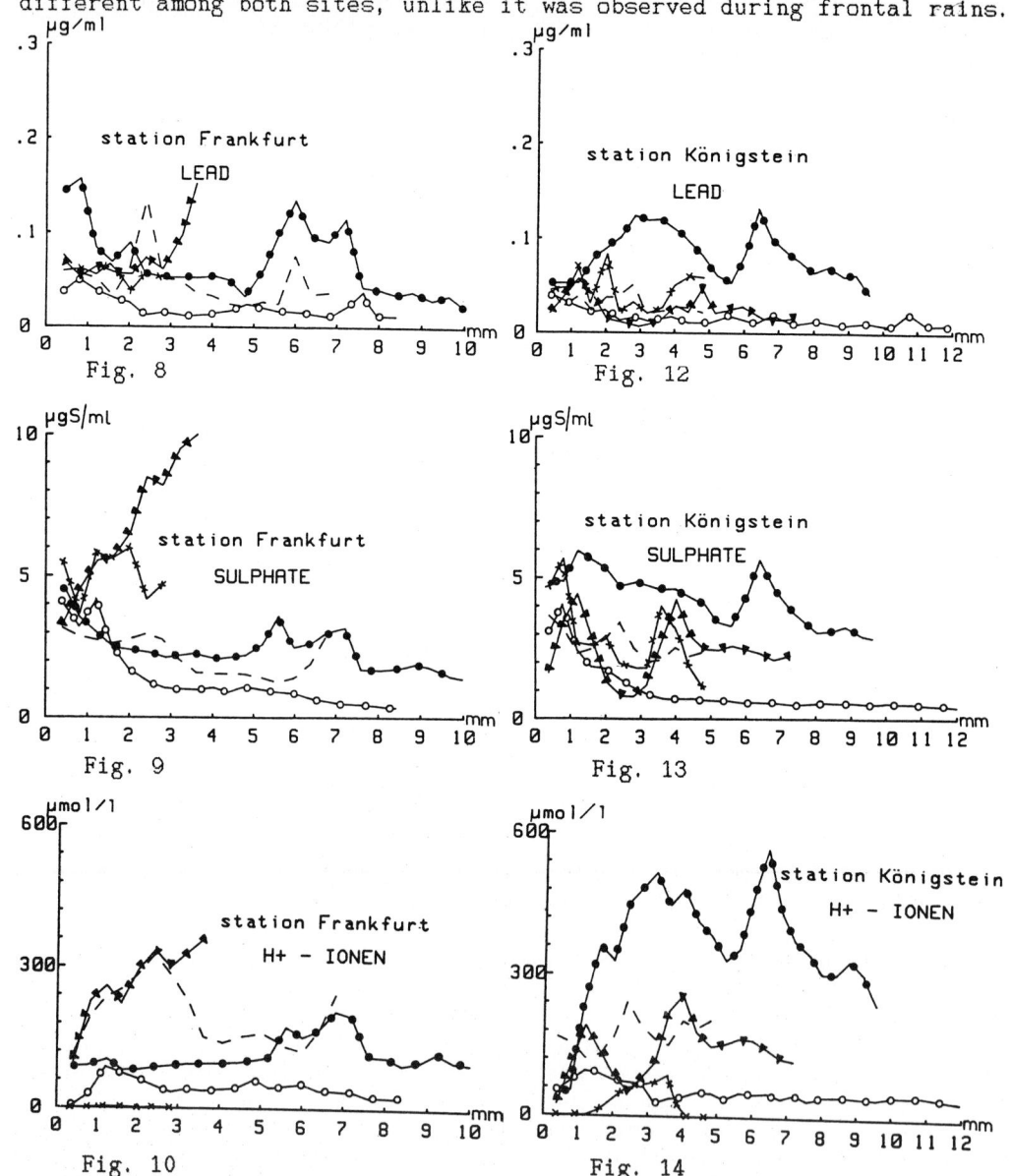

Fig. 8
Fig. 12
Fig. 9
Fig. 13
Fig. 10
Fig. 14

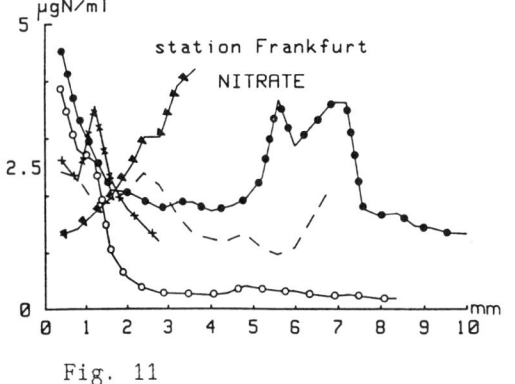

Fig. 8-14: concentration variations of Pb, $SO_4^=$, NO_3^-, and H^+-ions during five convective showers.

○—○ 5.5.84   — — 16.5.84
▼—▼ 27.5.84   ✕—✕ 11.7.84
●—● 24.8.84

Fig. 11

3.3. Contribution of rainout and washout in the wet deposition originating from frontal rains

The estimate is based on the assumption that the constant level of contaminants in rainwater which in frontal rains is approached after some time has passed is due to rainout only, since at this time airborne pollutants below the cloud have already been scavenged. The final five samples of each precipitation event were selected as being representative for this contribution of rainout. The rainout-fraction, given as the percentage-fraction which the final concentration constitutes of the corresponding initial concentration, was calculated for each of the compounds considered. The rainout-fraction for each compound is given in Fig. 15 both for the five individual cases analyzed and their arithmetic mean and standard deviation.

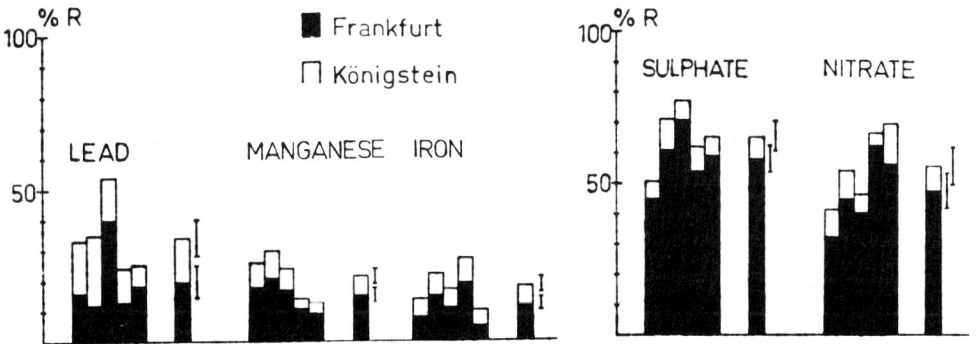

Fig. 15: percentage of rainout (R) with respect to total scavenging at Frankfurt and Königstein

The rainout-ratios were found to be higher at Königstein than at Frankfurt, for each compound analyzed and for each of the individual cases of rain considered. For the heavy metals analyzed, Pb has the largest rainout-fraction; i.e. it has the lowest contribution of below cloud scavenging. In all of the individual cases of rain sampled, the rainout-

ratios were larger for the anions than for the heavy metals. This is due
to the presence of gaseous precursors of the anions (e.g. SO_2, NO_x,
HCl), which are incorporated into the rain predominantly by in cloud
scavenging. Although the kinetics of absorption and oxidation of SO_2 and
NO_x in droplets is not completly understood and still under investigation, it seems likely that below cloud scavenging is not an efficient
mechanismus for these gases, due to the short time while the droplets
are falling towards the earth's surface (Castillo et al. 1983).

Effect of the evaporation of falling droplets on the concentration of
constituents in rainwater
Kühme (1968) has elaborated a scheme which describes the effect of evaporation of falling droplets on the droplet spectra as a function of
height, temperature and humidity. Applying Kühme's model, the contribution of evaporation to the final concentration in the rain sample was
estimated. For the five events considered here, evaporation was found to
be negligible, due to the low ceilings observed during the rain. Also,
the difference of 400m in altitude between the two sites does not
produce significant difference among the concentrations at both sites.
Thus, the difference between 100% and the rainout-fractions as given in
Fig. 15, can be interpreted as the contribution of washout to precipitation scavenging.

3.4. Soluble and insoluble metal compounds in rain

The insoluble fraction of a metal compound in rain is operationally
defined as the fraction which is retained on membrane filters.
For the metals analyzed, the partition among the insoluble/soluble
fractions was not found to vary significantly with the rainfall amount
or among the different sampling sites.
Among the different rains sampled, Fe and Mn show only minor variations
of the insoluble fraction (Fe: 75-99% insoluble, Mn: 18-34% insoluble),
whereas for Pb the variability is much larger (7-89% insoluble). Plotted
versus the corresponding pH in the samples (Fig. 16), the fraction of
soluble Pb obviously increases with decreasing pH, whereas the soluble
fractions of Fe and Mn show no distinct dependence from pH.
Rohbock (1984) suggests, that the insoluble Pb might be present as ions
which are adsorbed on particulates and are released as the concentration
of free acid increases.

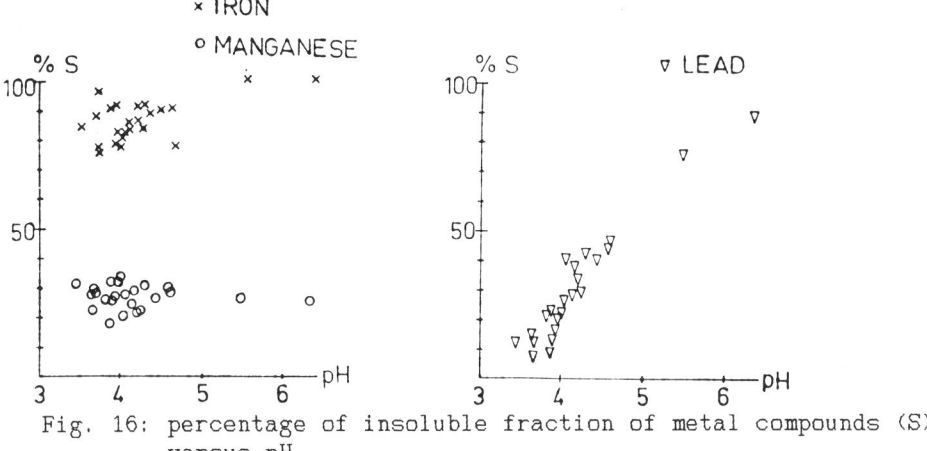

Fig. 16: percentage of insoluble fraction of metal compounds (S) versus pH

4. CONCLUSIONS

The analysis of sequential rain samples collected simultaneously at different sampling sites yields the following results:
1. The wet deposition of heavy metals was significantly higher at the polluted urban site than at the rural site. Anions were distributed more homogeneously, the difference among the sites was only minor for $SO_4^=$, NO_3^- and Cl^-.
2. In rain from stratified clouds the concentration of contaminants shows a very regular pattern: it peaks in the initial phase and subsequently declines towards the end. In convective precipitation the concentration scatters irregularly, as the precipitation intensity does.
3. For all the compounds analyzed, the contribution of below cloud scavenging to wet deposition was found to be largest at the urban site, as compared to the rural sites.
4. At all sites, the anions contributed a larger fraction to rainout than the heavy metals did.
5. The fraction of soluble Pb increases as the acidity of rainwater increases between pH 6.4 and pH 3.4, for Fe and Mn no such relationship is observed.

REFERENCES

Asman, W.A.H. (1980):'Draft, construction and operation of a sequential rain sampler' ; Water, Air, and Soil Pollution 13 pp 235-245

Castillo, R.A., Jiusto, J.E., McLaren, E. (1983):'The pH and ionic composition of stratiform cloud water'; Atm. Env. Vol.17 pp 1497-1505

Coscio, M.R., Pratt, G.C., Krupa, S.V. (1982):'An automatic, refrigerated, sequential precipitation sampler'; Atm. Env. Vol.16 pp 1939-1944

Dawson, G.A. (1978):'Ionic composition of rain during sixteen convective showers'; Atm. Env. Vol.12 pp 1991-1999

De Pena, R.G., Carlson, T.N., Takacs, J.F., Holian, J.O. (1984): 'Analysis of precipitation collected on a sequential basis'; Atm. Env. Vol.18 pp 2665-2670

Georgii, H.W., Perseke, C., Rohbock, E., (1982): Feststellung der Deposition von sauren und langzeitwirksamen Luftverunreinigungen aus Belastungsgebieten; Abschlußbericht zum Forschungsprojekt 104 02 600 im Auftrag des Umweltbundesamtes, März 1982

Kins, L. (1982): Differentialanalysen der chemischen Zusammensetzung von Einzelniederschlägen; Diplomarbeit, Inst. f. Meteorologie and Geophysik, Univ. Frankfurt/M

Kühme, H.W. (1968):'Ein Verfahren zur Bestimmung der Größenänderung fallender Regentropfen in Abhängigkeit von Temperatur und Feuchte'; Berichte des Inst. f. Meteorologie und Geophysik, Univ. Frankfurt/M Nr. 15

Martin, A., Barber, F.R. (1984):'Acid gases and acid rain monitored for over 5 years in rural east-central England'; Atm. Env. Vol.18 pp 1715-1724

Raynor, G.S., Hayes, J.V. (1983):'Differential rain and snow scavenging efficiency implied by ionic concentration differences in winter precipitation' in: Precipitation Scavenging, Dry Deposition, and Resuspension, Vol.1 Coordinators: Pruppacher, H.R., Semonin, R.G., Slinn, W.G.N. pp 249-264

Rohbock, E. (1984): Der atmosphärische Eintrag von Schwermetallen über trockene und feuchte Deposition; Dissertation, Inst. f. Meteorologie und Geophysik, Univ. Frankfurt/M

Seymour, M.D., Stout, T. (1983):'Observations on the chemical composition of rain using short sampling time during a single event'; Atm. Env. Vol.17 pp 1483-1487

PRELIMINARY RESULTS AND EXPERIENCES WITH A NEW IN-SITU
MEASUREMENT SYSTEM OF RAINFALL ACIDITY IN FOREST AREAS
OF RHEINLAND-PFALZ / FEDERAL REPUBLIC OF GERMANY

Horst Borchert

Landesamt für Umweltschutz und Gewerbeaufsicht, Rheinland-Pfalz, 65 Mainz, Rheinallee 97, Federal Republic of Germany

ABSTRACT

Acid depositions are supposed to be responsible, at least in part, for the increasing forest decline. To determine the real acidity of precipitation it is expedient to measure pH of the rain immediately after its falling, quasi in situ. In the forests of Rheinland-Pfalz four in situ precipitation monitors had been installed. They are telemetrically computer controlled by the measuring system ZIMEN. Preliminary results show similar developments of acidity in the rain during its falling at different measuring points in the country.

INTRODUCTION

In Rheinland-Pfalz/Federal Republic of Germany, which is one of the most forested regions of Central Europe, air-pollutants and meteorological parameters are monitored by the telemetrically controlled realtime system ZIMEN. ZIMEN is the German abbreviation of the word "Zentrales Immissionsmeßnetz". In towns with large industries sixteen monitoring stations are operating continuously to determine the longterm trend of air pollution and to control the short-term behavior of pollutants during smog situations (1). Furthermore, four measuring stations have been installed in the forests all over the country with distances between 50 and 130 Km from each other to investigate the causes of forest decline by measuring the gaseous pollutants SO_2, NO_2, NO, O_3, and Dust and the meteorological components wind direction, wind velocity, humidity, temperature and globally rays (2). At these stations, a system has been installed continuously monitoring pH-value, electrical conductivity, amount and intensity during precipitation events. The monitor is a wet only collector free from pertubations due to dry depositions. To guarantee that the in situ monitor is at any time able to analyse the rain, it is controlled by a computer, which is installed in the telemetrically controlled measuring station.

THE MEASURING SYSTEM

The principle of the measuring system has already been described by Winkler (1978,1984). The rain passes through a funnel, which is automatically opened by a rain sensor, into a small tube, which keeps the liquid at a temperature of 25°C. Then it runs through a cell, which measures the conductivity, and after this the water falls in drops into the pH measuring chamber, which is made of teflon. In this way, the conductivity measuring chamber and the pH measuring chamber are galvanometrically separated to avoid them influencing the measurements of each other. To measure the conductivity and the H^+ ion concentration commercially available measuring instruments are used. From the pH probe the water drops into the rain volume measuring system using the seesaw method. In this way the volume of rainfall is translated into electrical pulses equivalent to 0.05mm per pulse.

DATA PROCESSING

The electrical analog signals are converted into digital signals by ADC's and translated into realtime half hour averages by a microprocessor in the measuring station. These averages are related to the volume of the rain fallen during the half hour simultaneously. The measuring station process computer (pdp 11 type) stores the half hour averages of the pH values and the conductivity values, the volume of rain per half an hour, the minimum pH value during each half hour, and some status signals of the measuring system, for instance whether the lid of the collecting funnel is open or closed and whether the thermostatically controlled installation box is at the right temperature. Every eight hours the main computer in the head office of the measuring network ZIMEN in Mainz contacts every station by telephone to get the measured results and status information as well as the half hour averages (5).

CALIBRATION AND TELEMETRICAL CONTROL

The ZIMEN computer system allows a telemetrical control of the correct function of the measuring system. By a telemetrical command the lid of the collecting funnel can be opened and a test sample of water can be pumped into the measuring system. This water is artificial rainwater with pH 4.6 and a conductivity value of 40 µS/cm at 25°C (6). The test results can be seen directly on the display in the

A NEW IN-SITU MEASUREMENT SYSTEM OF RAINFALL ACIDITY IN FOREST AREAS

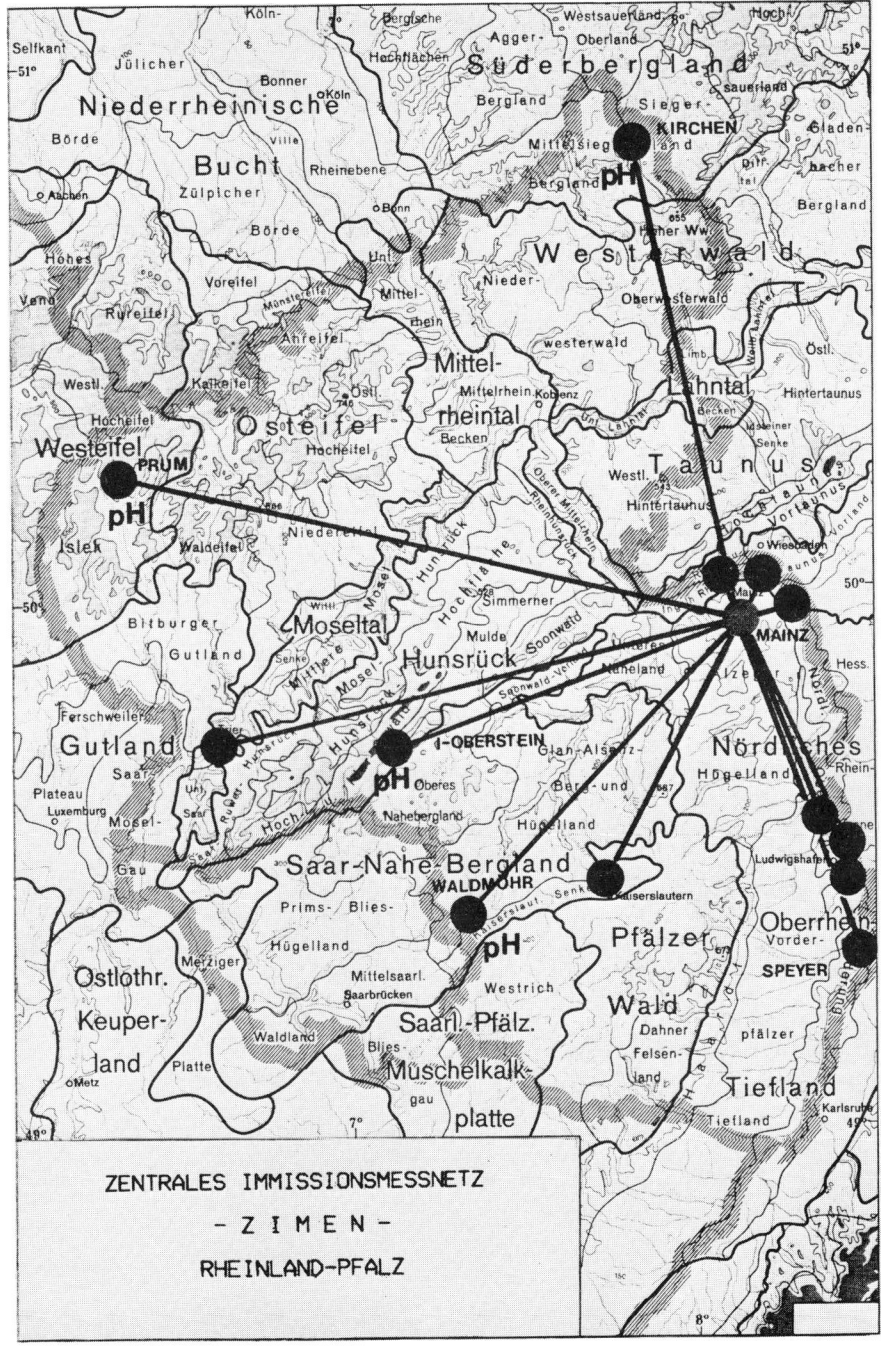

Meßobjekte: SO_2, NO_2, NO, O_3
Staub,
$C_nH_m - CH_4$, CH_4, CO;
Meteorologie

Fig. 1

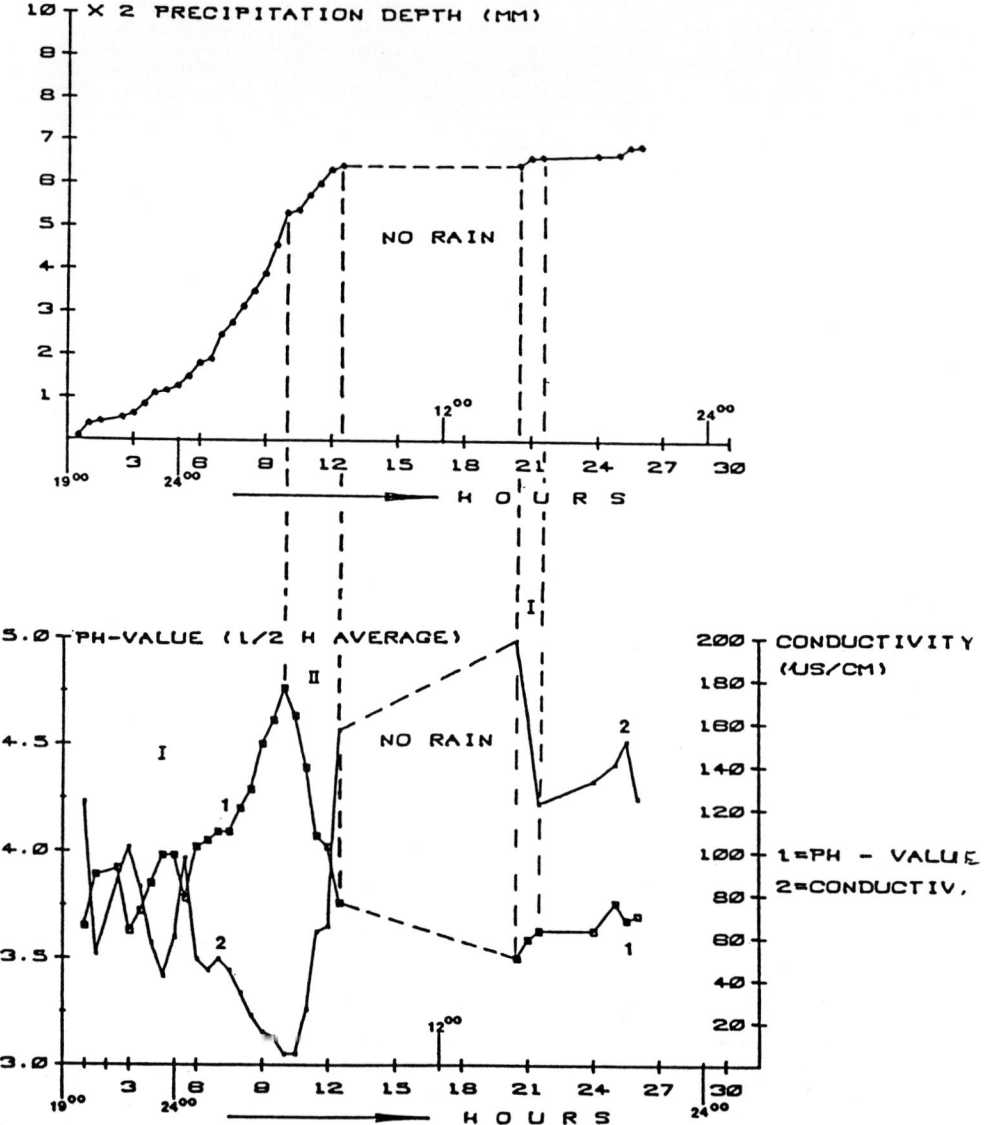

Fig.2

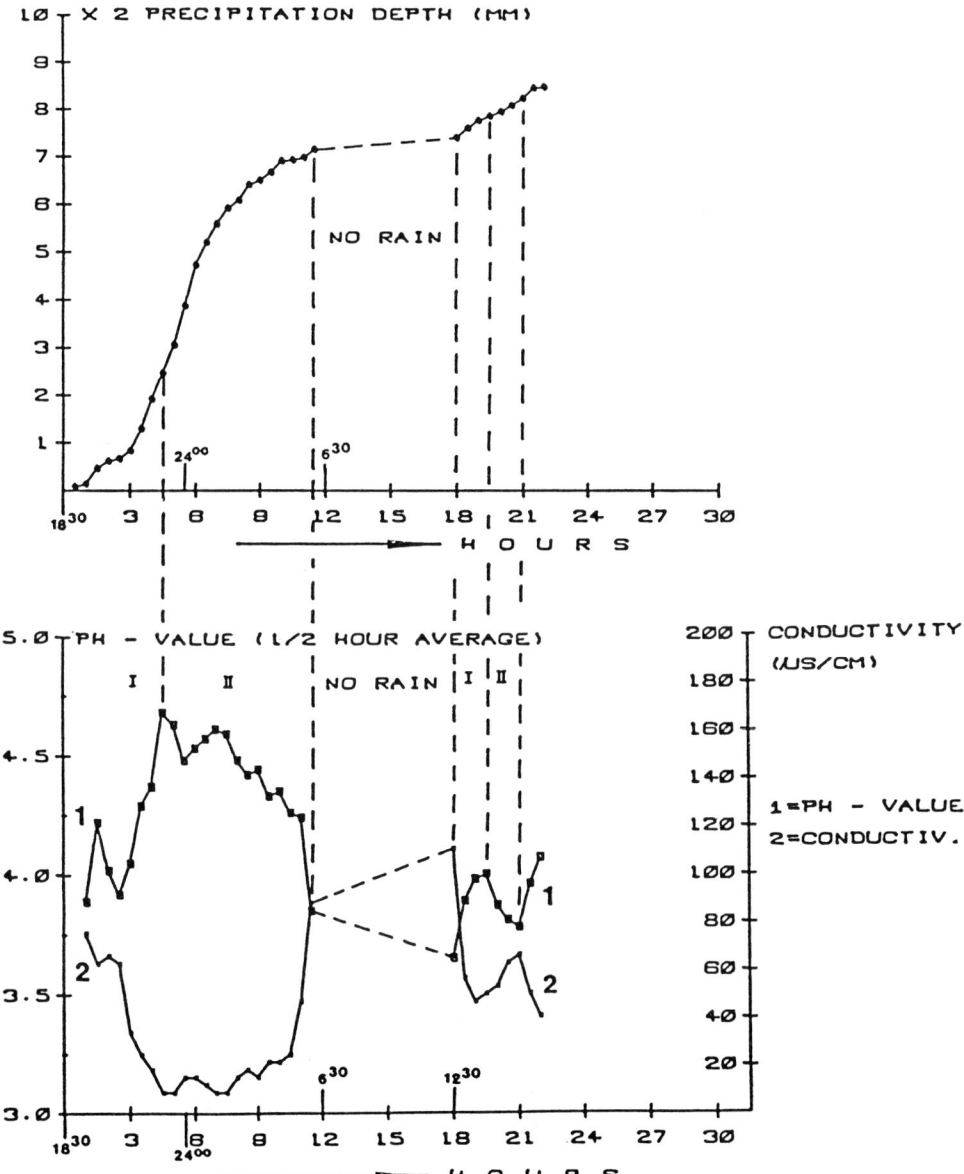

Fig.3

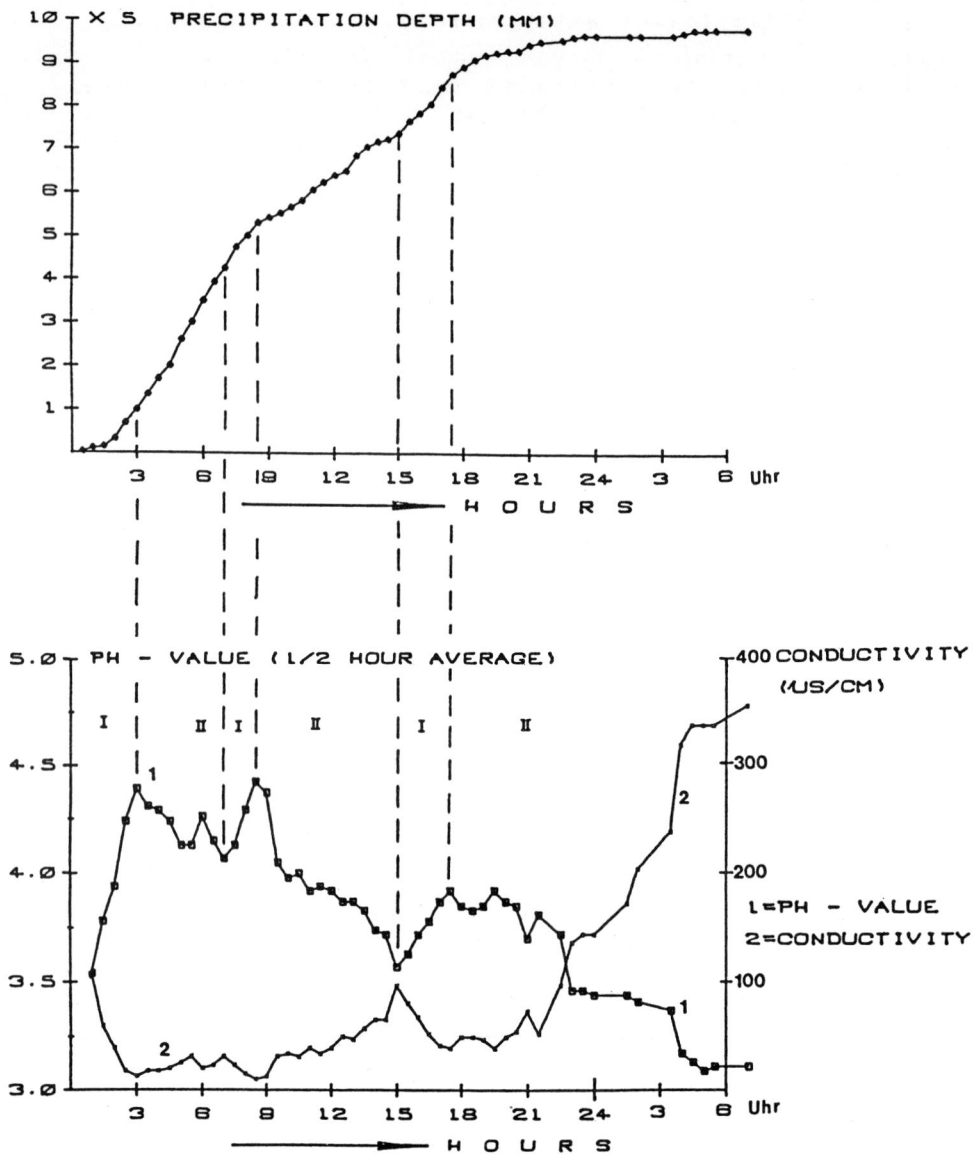

Fig.4

head office. In this way the performance of the measuring system can be controlled and the dehydration of the measuring system during long rainless periods can be avoided without having to visit the measuring station. In order to ensure long-term stability of the calibration function of the pH-probe we use the KCL-gel filled Schott electrode N 7781. The experience shows that it is sufficient to visit these telemetrically controlled measuring systems every 3 or 4 weeks. At each visit the system has to be cleaned. To calibrate the pH-electrode it is put outside the measuring chamber into two differend pH-calibration standard buffer solutions. Inside the chamber we control the pH-meter and the conductivity-meter with synthetic rain mentioned above.
The standard derivation of the conductivity values vary around $\pm$ 2 μS/cm, the standard derivation of the pH-values vary around $\pm$ 0.05 pH.

MEMORY EFFECT

The continuous measuring method has a basic problem which has to be considered concerning measurement errors, especially for small volumes of rain: When the rain starts falling, about 7ccm (that is equivalent about 0.15 mm) of the new rain is necessary to displace the old rain out of the conductivity probe, and correspondingly another 7 ccm are necessary to displace the old rain out of the pH-measuring-chamber. these memory effects should be taken into account, when the averages are calculated by the processor.

MEASUREMENTS AND CONCLUSIONS

Preliminary results show similar developments of acidity in the rain during its falling at three different measuring points in the country: From June 19 to June 22, 1985, we had a rain period with strong westerly winds and relatively cold weather. This period started simultaneously in the western stations at Idar-Oberstein/Hunsrück and Prüm/Eifel. (At that time in the station at Waldmohr/Westpfalz the rain monitor had not jet been installed). In the eastern station at Kirchen/Westerwald the rain began about 5 hours later. The upper graph in Fig.2 - 4 shows the cumulative rainfall amount during the whole period. The lower graph shows half hour averages of the pH values (left scale) and half hour averages of the conductivity values (right scale).
Within a rain period two different phases can be distinguished:
In phase I, which is at the beginning of the rain period or at the beginning of the rain after a longer rain interruption, the pH value increases, while the conductivity decreases. During this phase the air seems to be cleaned

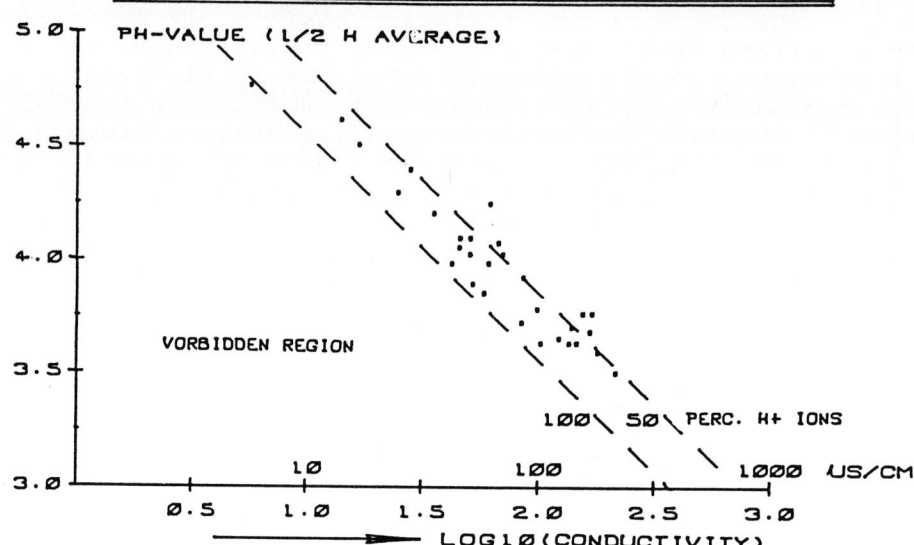

Fig.5

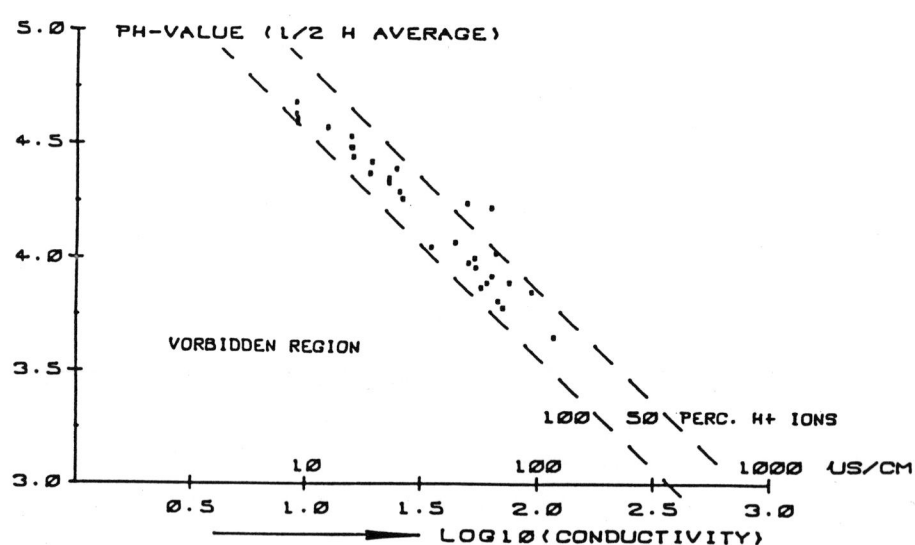

Fig.6

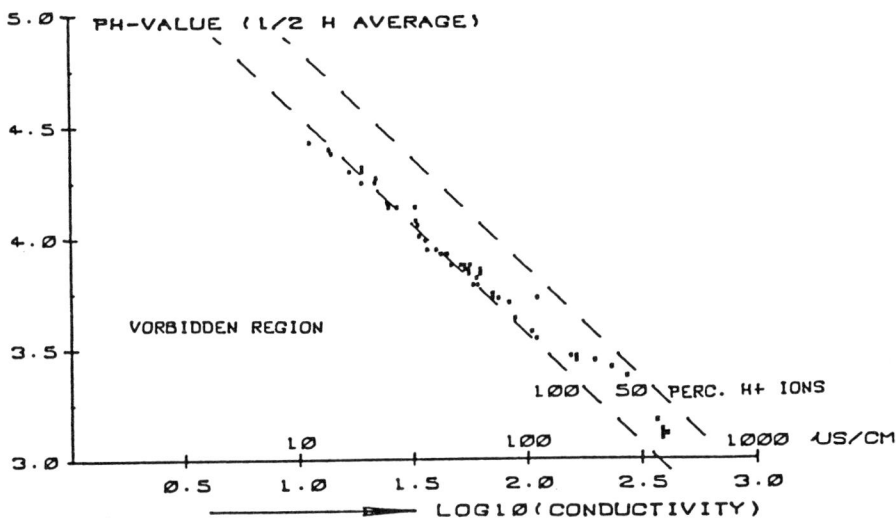

Fig.7

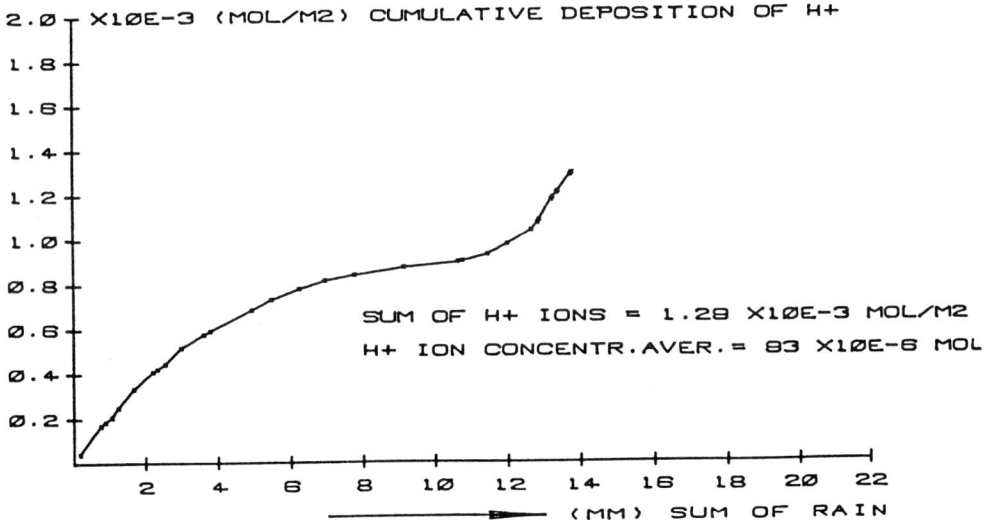

Fig.8

ZENTRALES IMMISSIONSMESSNETZ - ZIMEN - F. RHEINLAND-PFALZ

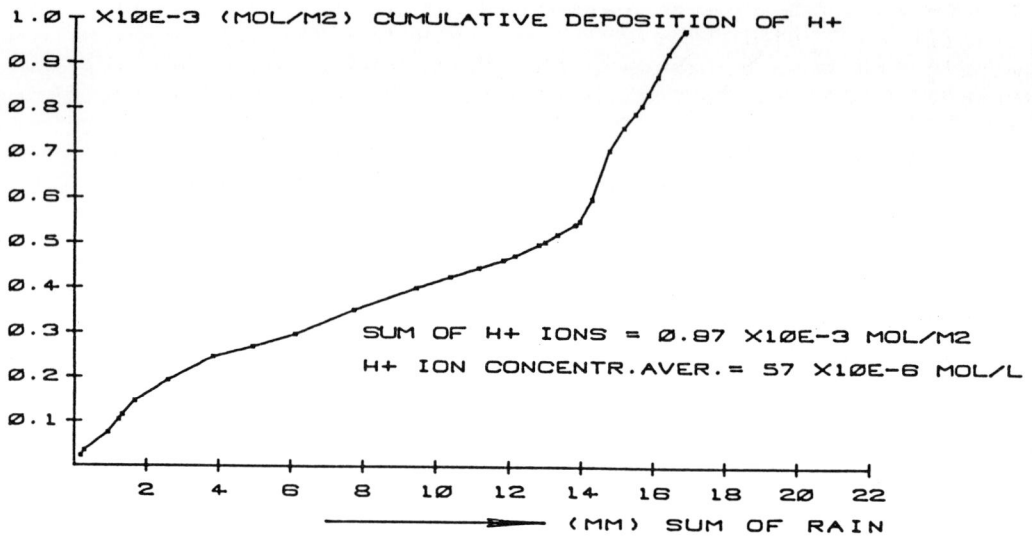

Fig.9

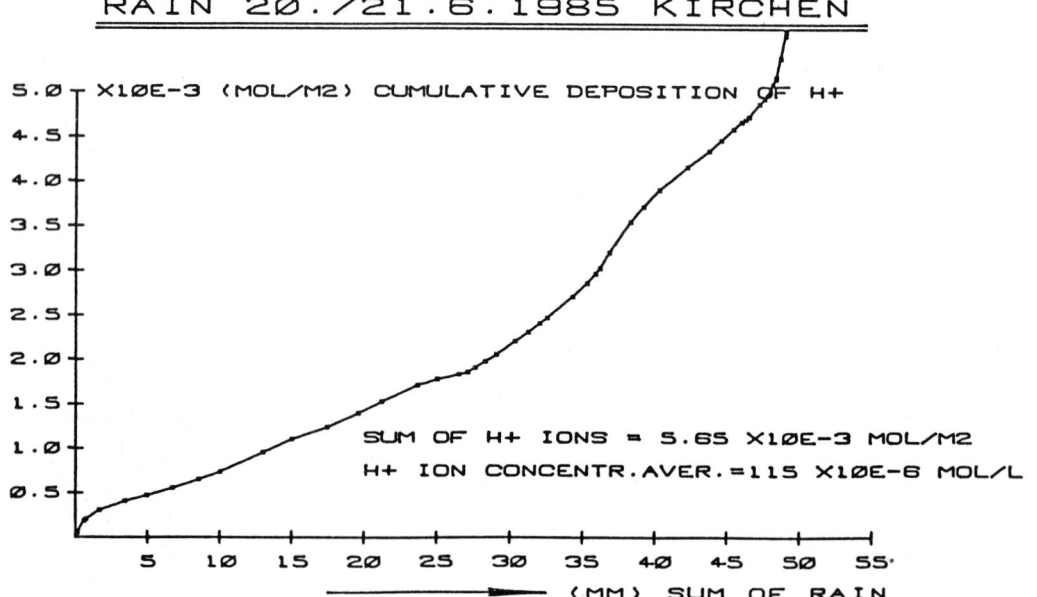

Fig.10

mainly from dust which is increasingly neutralizing disolved acids. The aerosol particles are scavenged more rapidly than acid can be produced (7).
In phase II we observe that the acidity in the rain increases with time. During this phase the velocity of wind mostly has been increased. The raindrops become smaller. Gaseous acids as SO_2 and NO_2 seem to be disolved during this phase increasingly.
The acidity of the rain depends on velosity, volume and temperature of the raindrops and on the solubility of the air pollutants (8).
To control the validity of the measured values they have to be compared with the theoretical graph of the relation between the pH value and the conductivity of extremely salt-free destilled water, which has the formula

$$Q = 5.544 - \log_{10}(L)$$

where   Q = pH value in pH units
        L = conductivity in $\mu S/cm$

Fig.5 to 7 show, that most of the measuring values are lying in the near of this graph, which is the border line of the "forbidden region" (4). The more these values lie in the righthand side of this graph, the more other ions than H^+ ions add to the conductivity of the rain.
A comparison of these plots of the three stations results, that in the western stations Idar-Oberstein and Prüm, which are lying near the industrialized regions of Saarland, Belgium and Lothringen, the fraction of non H^+ ions has been greater than in the station Kirchen lying about 130 Km eastwards.

ACID DEPOSITION

Fig.8 to 10 present the cumulative percentage of H^+ deposition. Especially during the last part of the rain period the acidity increased in spite of the decreasing intensity of the rain in all three measuring stations.
During the rain period from June 19 to 22, 1985, we measured an average H^+ ion concentration between 6 and 12 x 10^{-5} nol/l at three differend stations.
The total accumulated H^+ depositions during this precipitation period amounted in the western stations to about 1×10^{-3} mol/m^3, in the eastern station to 6×10^{-3} mol/m^3.

REFERENCES

(1) "Luftreinhaltung in Rheinland-Pfalz".(1978)
    Publisher: Landesamt für Umweltschutz und Gewerbeaufsicht, Rheinland-Pfalz, D65 Mainz, Rheinallee 97-101

(2) "Monatsberichte über die Meßergebnisse des Zentralen Immissionsmeßnetzes - ZIMEN - für Rheinland-Pfalz"
    Editors: Borchert, H. and Kampe, U.
    Publisher: Landesamt für Umweltschutz und Gewerbeaufsicht, Abt.3: Meßinstitut für Immissions-, Arbeits- und Strahlenschutz, Rheinallee 97-101, D65 Mainz, F.R.G.
    ISSN 0720-3937.

(3) Winkler, P.:(1978)
    "Fehler bei der Spurenstoffanalyse im atmosphärischen Niederschlag, dargestellt am Beispiel von pH-Wert und elektrischer Leitfähigkeit", Staub 38, pp 175-177.

(4) Winkler, P.:(1984)
    "The Precipitation Monitor"
    WMO, Instruments and Observing Methods Report Nr.15
    TELEMO Nordwigherhout 24./28. Sept.84, pp 115-119

(5) Borchert, H.:(1980)
    "Bedeutung und Anwendung von Richtlinien für Fernüberwachungssysteme im Umweltschutz"
    Real-Time Data Handling and Prozess Control.
    Editor: Meyer M., North-Holland Publishing Company, Brüssel

(6) "Monatsberichte aus dem Meßnetz" Nr.3, 1983 des Umweltbundesamtes(UBA), Berlin 33, Bismarckplatz 1:
    "Messungen von Leitfähigkeit und pH-Wert im Niederschlag".

(7) Winkler, P.:(1984)
    "Evaluation of the Information from a continuously working Precipitation Monitor".
    Third European Symposium, Varese (Italy),10./12.Apr.84, pp 467-472

(8) Beilke, S.:(1969)
    "Neue Ergebnisse über Auswaschung atmosphärischer Spurengase und Aerosole". Analen der Meteorologie, N.F. Nr.4, pp 122-125.

OCCURRENCE OF GASEOUS POLLUTANTS IN FOREST STANDS

Günter Baumbach
Institut für Verfahrenstechnik und
Dampfkesselwesen der Universität Stuttgart
Abteilung Reinhaltung der Luft
Pfaffenwaldring 23, D 7000 Stuttgart 80

ABSTRACT. As a contribution to the investigation of the causes of the forest-damage measurements of air pollution are carried out in the forest area "Schönbuch" 30 km south of Stuttgart and in the Black Forest near Freudenstadt. At the Schönbuch site the influence of the metropolitan area of Stuttgart can be clearly concluded from the nitrogen oxide immissions observed. In the Black Forest, mountain-valley-systems may lead to the accumulation of gaseous pollutants resulting in relatively high concentrations. There the influences of atmospheric barrier layers are of special significance. In the winter during east and northeast wind episodes long range transport of SO_2 is registered. Measurements of the concentration differences between above the canopy and in the forest stand provide information about the pollutant uptake by the trees. E.g. the concentration differences depend on the reactivity and the water solubility of the gases.

1. INTRODUCTION

As a contribution to the investigation of the causes of forest-damages measurements of air-pollution are carried out by our institute in two forest areas in Baden-Württemberg. Measuring station 1 is situated in the forest area "Schönbuch" 30 km south of Stuttgart at a height of 500 m above sea-level. Station 2 lies near Freudenstadt in the northern Black Forest at a height of 800 m above sea-level. In both stations the gaseous pollutants sulfur dioxide (SO_2), nitrogen oxide (NO), nitrogen dioxide (NO_2) and ozone (O_3) as well as meteorological parameters like wind direction, wind speed, temperatures, solar radiation and further data are continuously measured. From these data half-hour mean values are calculated. Not only the occurance of gaseous pollutants in the forests is registered but also statements about the immission of gaseous pollutants can also be made by measuring the differences above the canopy and in the forest stand. Figure 1 shows the arrangement of such a forest measuring station. The measuring techniques are described in detail elsewhere /1/. The measuring station in the Schönbuch Forest has been operating since December 1983, the station near Freudenstadt since June 1984. Here some of the results are presented.

Figure 1. Arrangement of a measuring station

2. AIR POLLUTION IN THE SCHÖNBUCH FOREST

Generally when the region is influenced by low pressure systems with westerly winds only very low concentrations of gaseous pollutants can be measured in the Schönbuch as well as in the northern Black Forest. In autumn and winter, however, especially during good-weather periods, increased NO_x concentrations are often registrated in the Schönbuch. As an example Figure 2a shows the variation of the NO_2 concentration of November 1984 (each half-hour value is shown as a vertical column). Figure 2b shows the corresponding ozone concentrations. In November 1984 the weather situation was mainly marked by continuous high pressure. Often relatively high NO_2 concentrations were measured in the forest area. The O_3 concentration data are in a remarkable inverse correlation to NO_2 data. Ozone is consumed by the oxidation of the nitrogen oxides. To evaluate the origin of the nitrogen oxides each half-hour mean value was assigned to the respective sector of the wind direction and the mean values for each sector calculated. This results in pollutant wind roses.

Figure 3 shows the pollutant wind roses of the components NO and NO_2 at the Schönbuch measuring station for the period from autumn to winter 1984. The increase of the nitrogen oxide concentrations in case of northerly winds can be seen quite distinctly. Clearly, this points to the influence of the metropolitan area of Stuttgart with its center 30 km north. Therefore, nitrogen oxides observed at the Schönbuch site predominantly originate in the Stuttgart area. During good weather periods Stuttgart itself is often covered by a brown veil of haze which indicates the accumulation of pollutants. Also the SO_2 concentrations in the Schönbuch Forest reflect the influence of Stuttgart as can be seen from Figure 3c.

In autumn and winter ozone is consumed by nitrogen oxides. The pollutant wind rose of the autumn-winter period as described in Figure 4a shows that only very low ozone concentrations exist in the wind directions where the nitrogen oxides are increased. The highest concentrations are registered with winds from southwest. In this direction there are no specific emission sources. The situation changes remarkably in spring and summer with increasing solar irradiation: The ozone wind rose for such a period, see Figure 4b, shows high concentrations with winds coming from the north and east although the emission sources are lying in this direction. This is certainly due to different heights of the mixing layer in autumn and late spring. In case of increased solar irradiation the photochemical conversion of the nitrogen oxides and hydrocarbons, emitted in the Stuttgart area, may, however, also contribute to the increased ozone-concentrations in the Schönbuch Forest area.

Whereas the increased NO_x concentrations are to be traced back to the influence of Stuttgart, the highest SO_2 concentrations are registrated generally with eastern and northeastern winds. Such a situation existed for instance on February 10th and 11th, 1983. In Figure 5a one can recognize how in those days the SO_2 was also measured by the higher situated measuring stations of the

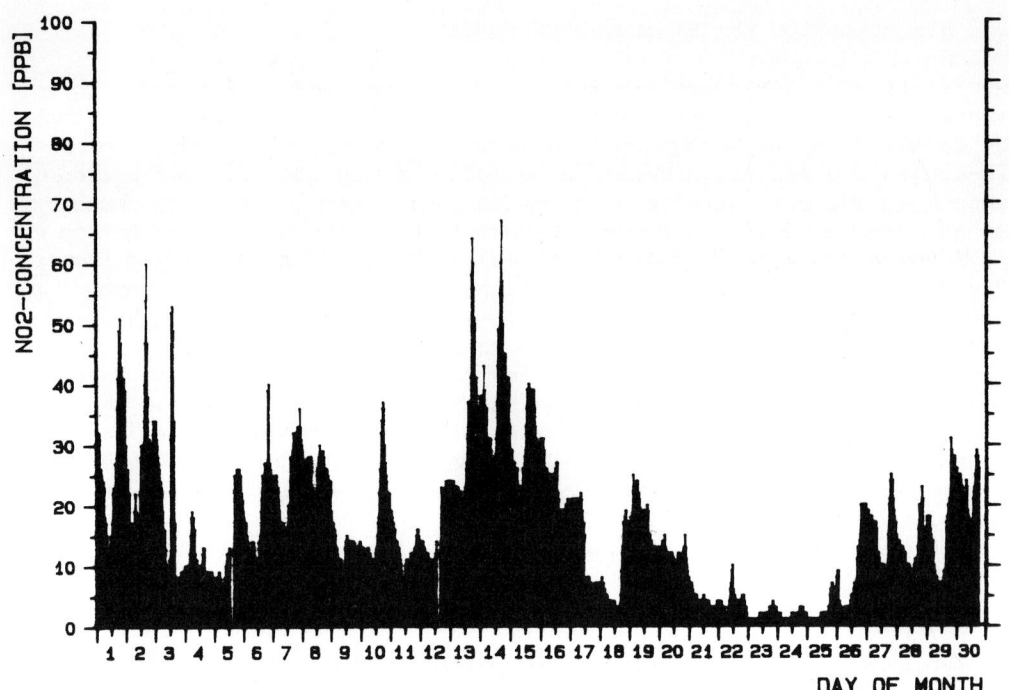

2a. NO₂ concentrations

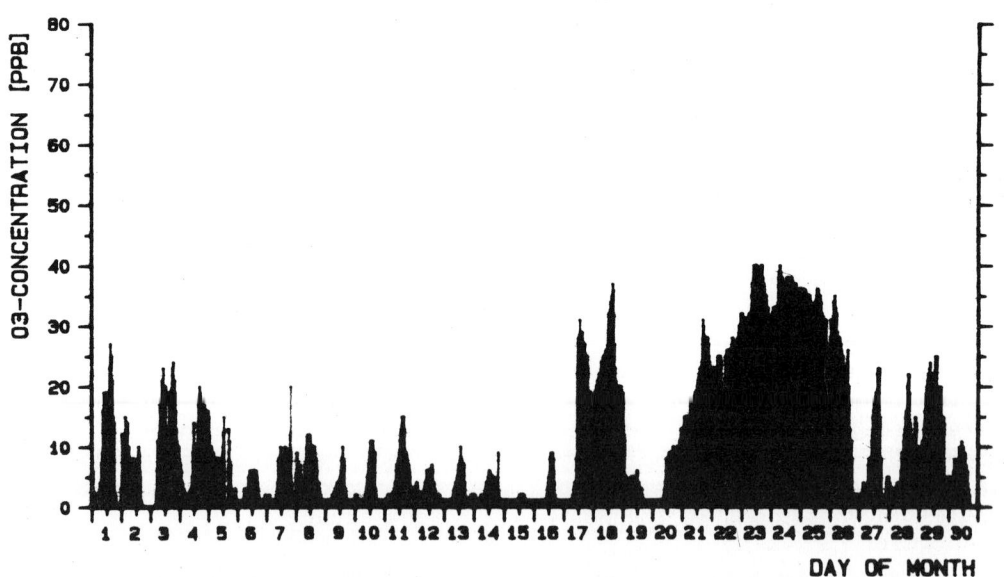

2b. O₃ concentrations

Figure 2. Concentrations of NO_2 and O_3 in November 1984; air quality measuring station in the Schönbuch Forest; each half-hour value is shown as a vertical column.

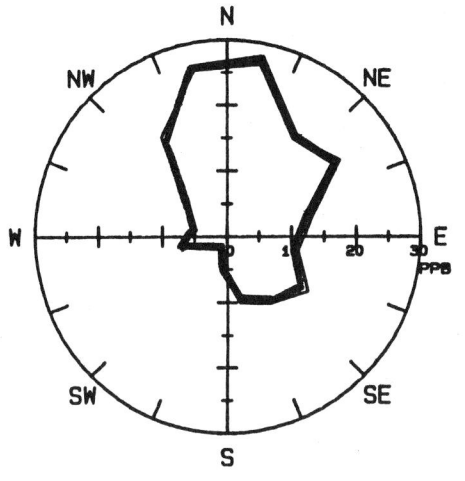

a. NO
Calms: 17,2 ppb above,
17,4 ppb below

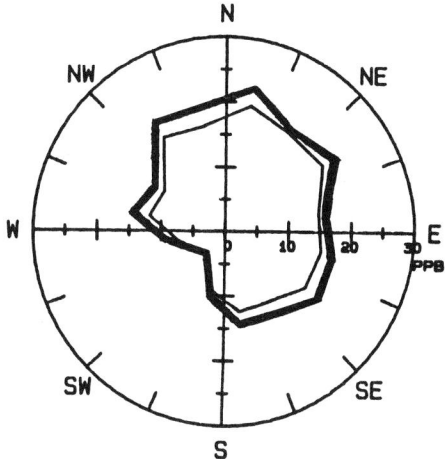

b. NO_2
Calms: 16,9 ppb above,
14,8 ppb below

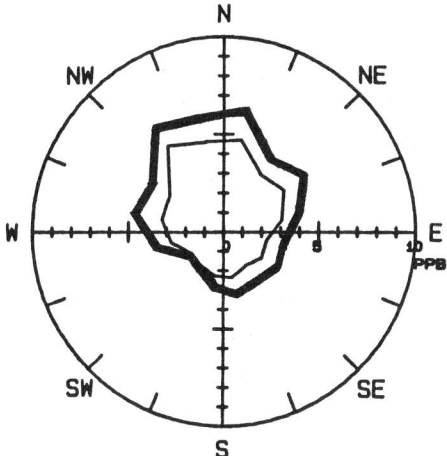

c. SO_2
Calms: 4,6 ppb above,
3,4 ppb below

Figure 3. Pollutant wind roses at the Schönbuch measuring station for the period from autumn to winter 1984 (average values)
━ above the canopy, — below the canopy

"Umweltbundesamt" in Southern Germany as shown in Figure 5b /2/. Here they are caused by large area long range transports probably originating from neighbouring countries in the East. Similar long range transport conditions were observed in December 1983 in Hamburg and Northern Germany /3/ as well as in January 1985 in the northern Black Forest /4/.

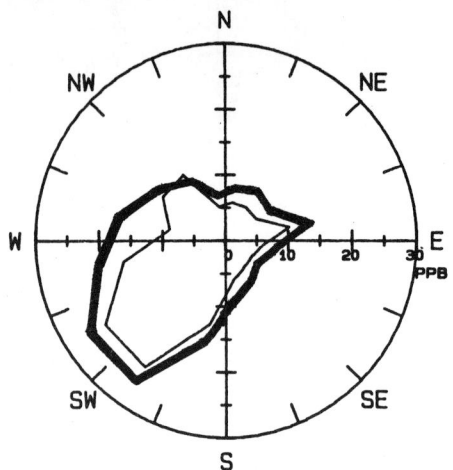
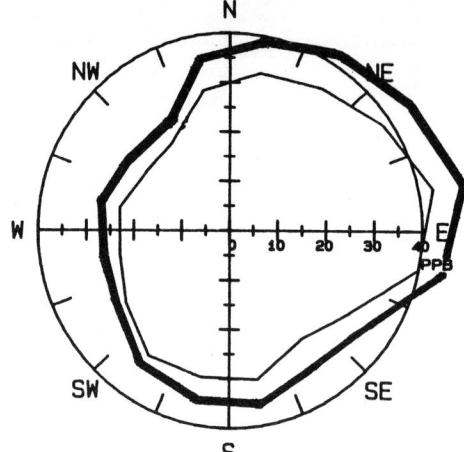

a. autumn winter
    period 1984
    Calms: 7,9 ppb above,
           6,1 ppb below

b. spring summer
    period 1985
    Calms: 27,7 ppb above,
           23,2 ppb below

Figure 4. Ozone wind roses at the Schönbuch station (average values)
▬ above the canopy, — below the canopy

3. AIR POLLUTION IN FOREST STANDS OF THE NORTHERN BLACK FOREST

The measuring station in the northern Black Forest was first installed in a forest region in the north of Freudenstadt. At this location especially during the night SO_2 peaks as high as 300 µg/m^3 combined with increased NO concentrations were often observed. The evaluation of the SO_2 wind rose (Figure 6) revealed that the SO_2 originated from the northwest. The emission source which caused these SO_2 peaks was discovered in a neighbouring valley: a heavy oil power plant with 28 MW thermal capacity, i.e. not a big installation. Control measurements in the valley showed that on slopes SO_2 half-hour values of 300 µg/m^3 up to more than 1000 µg/m^3 are often occuring. In mountain-valley systems obviously very high gaseous pollutant concentrations are occuring caused by atmospheric barrier layers. This was reported separately /5,6/.

Meanwhile the measuring station was transferred to a hill, 840 m above sea-level, situated in the south of Freudenstadt. At this location till now no significant emittents nearby were detected. The pollutant wind roses are balanced.

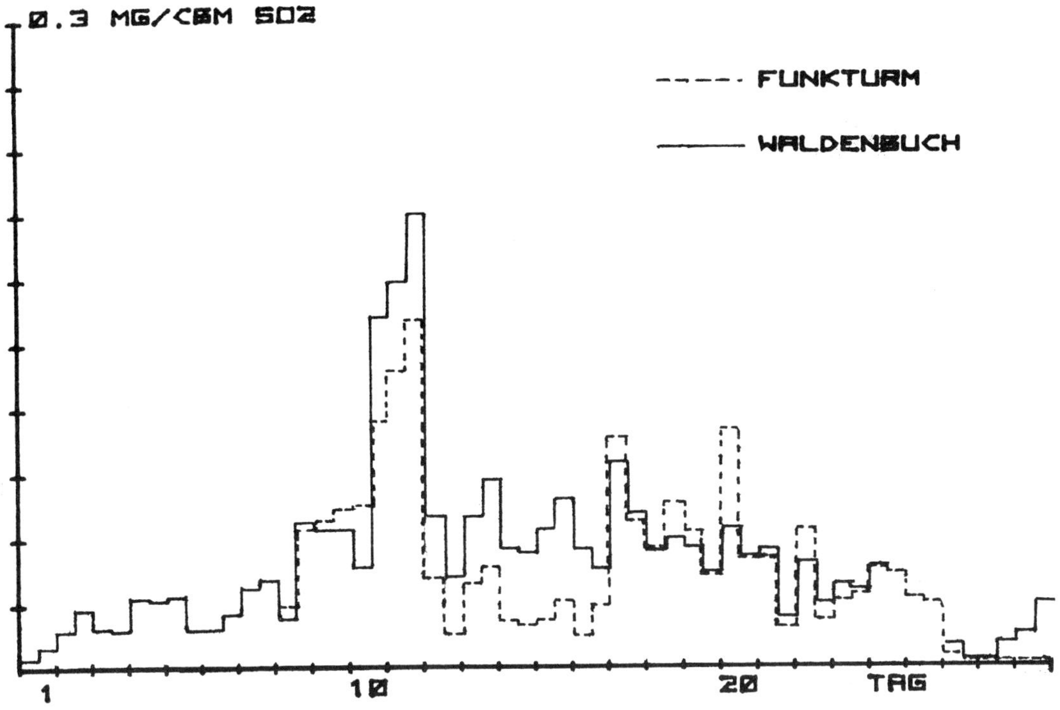

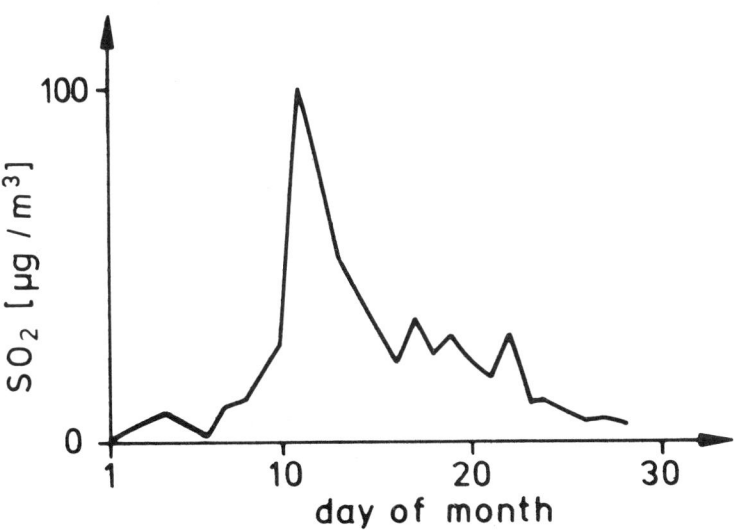

a. Variation at several measuring stations in the Schönbuch

b. Variation at the measuring stations of the "Umweltbundesamt" on mountains in Southern Germany

Figure 5. SO_2-concentrations in February 1983

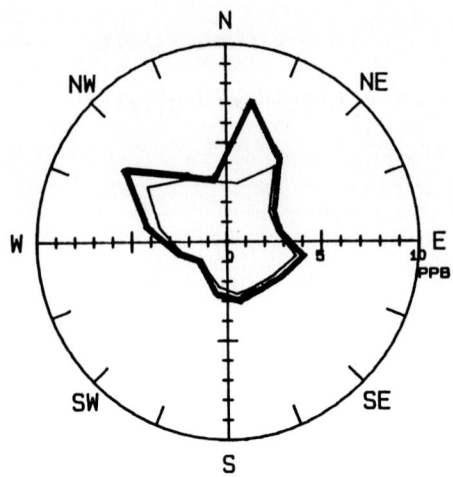 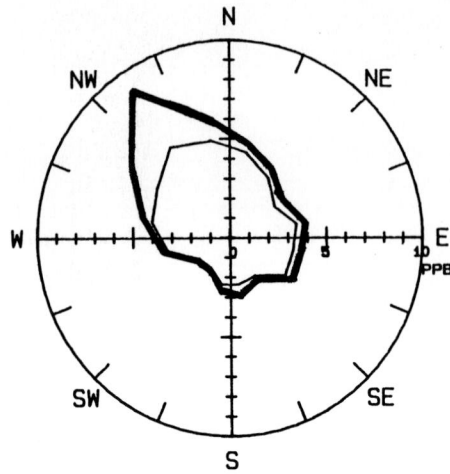

a. day time, 6.30 - 18 Uhr
Calms: 3,6 ppb above
3,3 ppb below

b. night time, 18.30 - 6 Uhr
Calms: 4,0 ppb above
3,4 ppb below

Figure 6. SO_2 wind roses at the first Black Forest measuring station (average values), ▬ above the canopy, —— below the canopy

During extended fine weather periods as in October 1985 an increase of the NO_2 concentrations is noticed regularly in the evening. The concentrations remain high until night. In Figure 7 this concentration variation is shown as the average diurnal cycle for 9 comparable days between the 16th and the 28th October 1985. Again O_3 and NO_2 concentrations behave oppositely. The SO_2 concentrations are very low and NO is near the detection limit.

The increase of NO_2 in the evening can be explained by the behaviour of the atmospheric barrier layers: Lowering of the elevated inversion layer and development of the ground inversion layer. For more detailed informations see /6/. In autumn, e.g. the atmospheric main barrier layers are specially moving in the middle altitudes of the low mountain ranges (800 to 1200 m above see level). Thereby increased pollutant concentrations can occur in those altitudes depending on the type of pollutant sources present in the area.

In the barrier layers also particles are concentrated. These accumulate gaseous pollutants by adsorption and serve as condensation nuclei. Fog droplets developed that way presumably contain especially high pollutant concentrations. Therefore fog measurements are particularly important for the determination of the immission situation in low mountain ranges. Depending on the weather situation varying pollutant concentrations might occur in the fog. This makes a continuous fog analysis highly desireable.

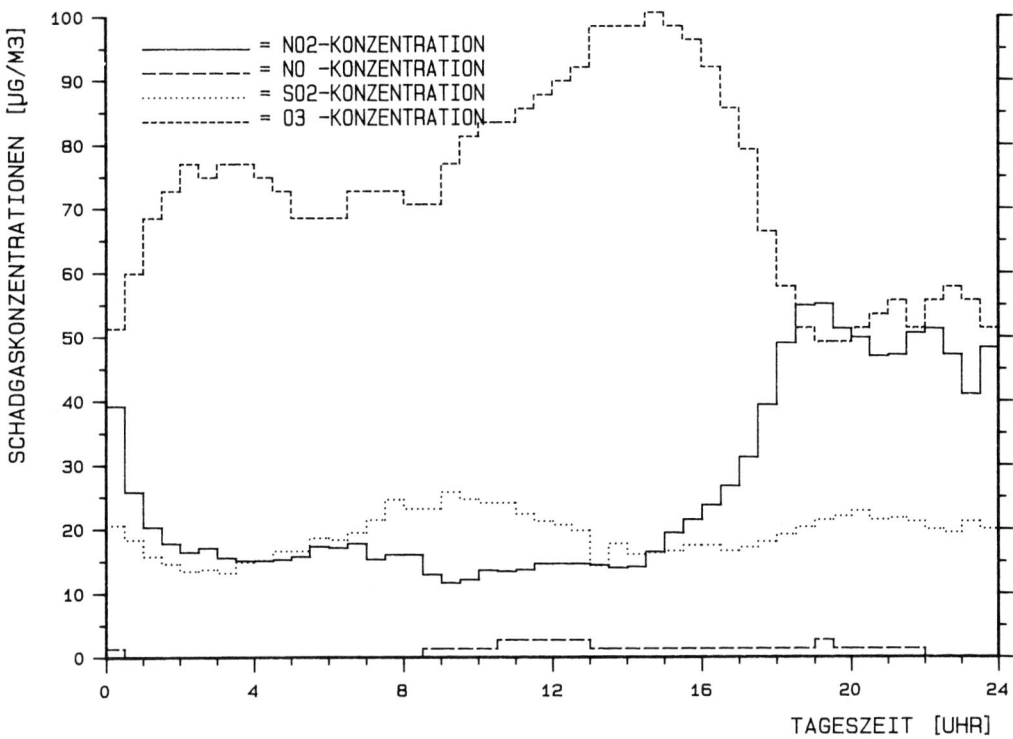

Figure 7. Average diurnal cycle of NO_2, NO, SO_2 and O_3 concentrations for 9 comparable days from 16th to 28th October 1985; measuring station in the northern Black Forest near Freudenstadt

4. DIFFERENCES IN GASEOUS POLLUTANT CONCENTRATIONS ABOVE AND BELOW THE CANOPY IN FOREST STANDS

The average values of the concentrations of gaseous pollutants measured above the canopy as well as in the forest stand are marked in the pollutant wind roses as seen in figure 3,4 and 6. There are two main causes of the observed concentration differences: Firstly, gaseous pollutants contained in the advected air are taken up by the trees, respectively are absorbed on the surfaces. Secondly, air passing over the forest does not penetrate the forest canopy. The absorption of gaseous pollutants depends on the one hand on the reactivity, water solubility and concentration of the gas and on the other hand on the absorption conditions of the trees. Reactive and water-soluble gases - SO_2, O_3 and also NO_2 - show higher concentration differences as for example NO with a low solubility in water.

The concentration differences also depend on the concentration level. The higher the concentrations the bigger are in most cases the concentration differences. In case of SO_2 the differences are most distinct when the trees are wetted.

In case of NO sometimes, specially during night, slightly higher concentrations are measured in the forest stand than above the canopy. This points to NO emissions from the soil. Compared with the transported quantities of NO however, the fraction emitted from the soil is small, see figure 3. The parameters of the concentration differences are presently under investigation.

5. SUMMARY

As a contribution to a better understanding of the causes of the progressing forest desease measurements of gaseous pollutants are carried out in forest stands of the Schönbuch (between Stuttgart and Tübingen) and in the northern Black Forest near Freudenstadt. The presence of the nitrogen oxides in the Schönbuch can undoubtedly be traced back to the influence of the metropolitan area of Stuttgart. Initially, considerably high SO_2 concentration peaks were measured in the Black Forest. They could be traced back to an emission source in a neighbouring valley. In mountain-valley systems very high gaseous pollutant concentrations can occur due to the formation of atmospheric barrier layers. Meanwhile the measuring station in the Black Forest was moved to a location where hardly any influences of nearby emittents can be detected. Also there the NO_2 concentration variations showed the influence of the atmospheric barrier layers.

Knowledge about the absorption of gaseous pollutants by trees can be gained by measuring the concentration differences above and below the canopy. It became obvious that the concentration differences depend on the reactivity of the gas and on its concentration. Significant differences between the concentrations of SO_2 above and below the canopy are measured especially in case of wetted forest stands. A minor NO emission from the soil could be observed as a result of differential measurements.

For further investigation of the forest damage it is necessary that in co-operation with investigations of biological effects the question is persued in which way gaseous pollutants are detrimental to plants. The answers to these questions have direct effects on the measuring techniques. For instance, wether the concentration, the dosis or the deposition of gaseous pollutants should be measured? Furthermore, which other parameters, for example photosynthesis-activity, radiation intensity, temperature, humidity or wind determine the extent of damage in case of direct immission of gaseous pollutants.

REFERENCES

/1/ Baumbach, G., Käß, M. (1985):'Ermittlung von vertikalen Schadstoffkonzentrations-Profilen in Waldbeständen in Baden-Württemberg - Meßtechnik und erste Ergebnisse' Staub-Reinhalt. Luft 45, Nr. 6, pp 274

/2/ Umweltbundesamt (1983): 'Monatsberichte aus dem Meßnetz' Nr. 2/83, Berlin, 1983

/3/ Bruckmann, P., Reich, T., Schrader, W. (1985): 'Die Hamburger Smogepisode im Dezember 1983' Staub-Reinhalt. Luft 45, Nr. 6, pp 307

/4/ Schweizer, G. (1985): 'Die Smog-Lage im Januar 1985 - Auswirkungen in Mittelbaden' Staub-Reinhalt. Luft 45, Nr. 12, pp. 587

/5/ Baumbach, G., Minner, G., Konrad, G. (1986): 'Luftverunreinigungen in einem Schwarzwaldtal bei Inversionswetterlagen' Bericht Nr. 3/1986, IVD-Abt. Reinhaltung der Luft, Universität Stuttgart

/6/ Baumbach, G.: 'Auftreten von Luftverunreinigungen in Mittelgebirgen durch Anreicherungen unter atmosphärischen Sperrschichten' report about the IMA-Querschnittsseminar "Atmosphärische Prozesse" 22./24.1.1986, Umweltbundesamt Berlin

POSSIBLE INFLUENCES OF CHLORINE CONTAINING SPECIES ON FORESTS

G. Fuchs, K. Bächmann
Fachbereich Anorganische Chemie und Kernchemie
Technische Hochschule Darmstadt
Hochschulstr. 4
D-6100 Darmstadt

ABSTRACT Some aspects of the behaviour of chlorine bearing coumpounds in the atmosphere are discussed. Examples of field measurements of particulate $\overline{Cl}$, gaseous HCl, chlorinated hydrocarbons and phosgene are presented and an estimation of the tropospheric Cl radical concentration is given.

1. Introduction
The possible reasons for the increasing forest desease in Western Germany can be classified into three types:
    1. Effects of air pollutants on needles and leaves (trace gases, aerosols, precipitation),
    2. effects of soil acidification and uptake of phytotoxic compounds by the plant roots, and
    3. microbial effects

A combination of all these effects may work together, although attacks of air pollutants seem to produce the major effects in all probability. Depending on the site considered also other effects may dominate.

    Many chemical systems in the atmosphere are possible to cause plant effects, they have however to fulfill some of the following conditions:
    1. Increase of immission concentration and/or source strengths,
    2. photochemical processes should participate because of the spatial distribution of the forest disease,
    3. some species of the system must have tropospheric lifetimes long enough to be transported to remote areas,
    4. effects on vegetation should occur in chamber experiments, and
    5. the effecting species or their degradation /reaction products should be detectable in needles and leaves.

The system of the chlorine bearing species in the atmosphere fulfills some of these conditions. Until now, it was not intensively investigated.

    Fig. 1 presents schematically the main reactions of chlorine bearing species in the atmosphere.

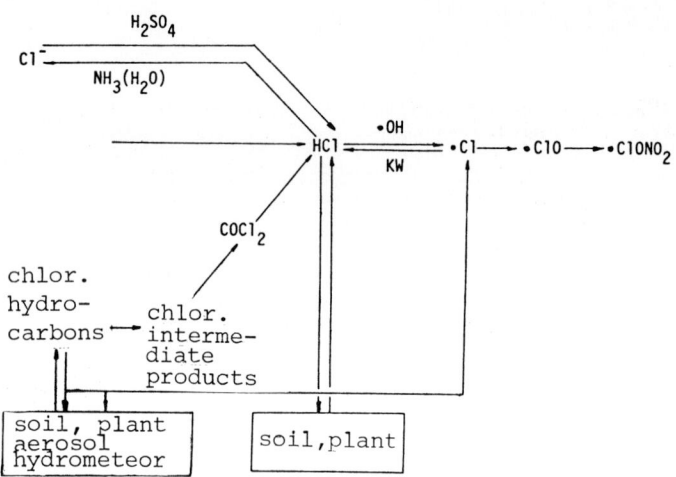

Figure 1. Main reactions of chlorine bearing species in the atmosphere

The importance of chlorinated hydrocarbons was first mentioned by Frank (1). We made field measurements of the inorganic chlorocompounds HCl and Cl^- (2-6). Zetzsch (7) investigated these species in a smog chamber.

The main chlorine bearing species in the atmosphere are particulate chloride, gaseous hydrogen chloride and chlorinated hydrocarbons. Particulate chloride is in equilibrium with gaseous HCl. This equilibrium is determined by the concentration of gaseous ammonia and the H^+ content of the particles.

In regions influenced by maritime air, the concentration of Cl^- is very much higher than in continental regions. The mean HCl concentrations are about 1 µg m^{-3} without any spatial dependence.

A second way to produce gaseous HCl is the photochemical degradation of chlorinated hydrocarbons, other degration products (besides HCl) are phosgene and choroacetyl chlorides.

By a reaction of HCl with OH radicals, Cl atoms can be formed which may form ClO and $ClONO_2$.

Cl radicals are álsoproduced by photochemical sensibilisation (8) e.g. at vegetation surfaces, from chlorinated hydrocarbons.

The following three tables compare the systems HCl/Cl^-, H_2SO_4/SO_4^{2-} and HNO_3/NO_3^-. Table 1 presents mean concentrations of the gasous species HCl and HNO_3 (H_2SO_4 does not exist in the gaseous phase). In Tab. 2 the anion content of aerosols and in Tab. 3 the anion content of rainwater is shown. The urban values were measured in Frankfurt, the rural values at a background station in Deuselbach in the German Hunsrück.

All values are calculated on a molar base.

Table 1: Anorganic gases (acids) in the troposphere (9,10)

	HCl	HNO_3
maritime	16	13
continental	27	56
above the inversion layer	3	3

(data in n mol m^{-3})

Table 2: Concentrations of anions in aerosols (11)
(mean values of 1981)

	Cl^-	NO_3^-	SO_4^{2-}
urban (Frankfurt)	25	86	113
rural (Deuselbach)	8	71	88

(data in n mol m^{-3})

Table 3: Concentrations of anions in rain water (11)
(mean values of 1981)

	Cl^-	NO_3^-	SO_4^{2-}
urban (Frankfurt)	32	54	59
rural (Deuselbach)	22	36	39

(data in n mol l^{-1})

Comparing the ratios Cl^-/NO_3^- and Cl^-/SO_4^{2-} in aerosols and in rain water, the ratios in rain water are higher than in aerosols. Four different explanations are possible for this increase:
1. Gaseous HCl is additionally dissolved in rain water,
2. most chlorine bearing salts are more soluble than sulfate bearing salts,
3. the amount of anions dissolved in rain water is due to the concentrations in the cloud (rain out) and the concentrations of the total column below the cloud (for interpretation, the vertical distribution of the considered species is needed), the wash out effect, and
4. wash out effects are dependent on the size distribution of aerosols.

2. Wash out of gaseous HCl

Fig. 2 presents the diurnal variation of the HCl concentration. At 2.40 p.m., it began to rain. The HCl concentration decreases correlated with the relative humidity.

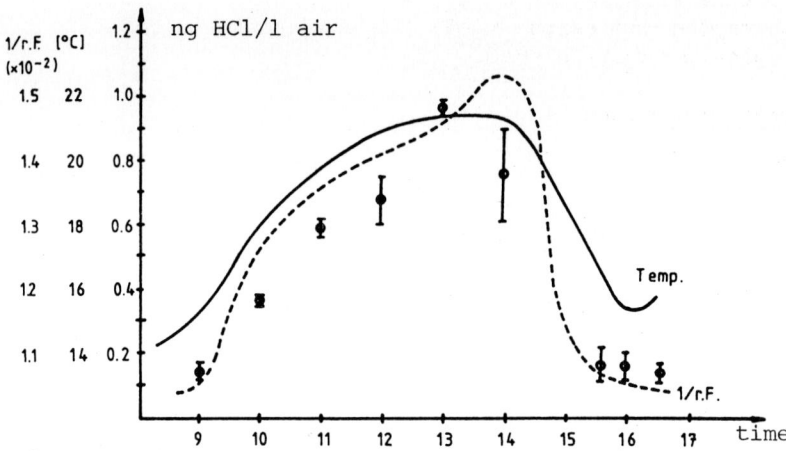

Figure 2: Diurnal variation of HCl concentration (9)

3. Vertical distribution of particulate chloride

Figure 3 shows an example of the vertical distribution of Cl^-, NO_3^- and SO_4^{2-}. In higher altitudes, the Cl^- concentration is higher than the NO_3^- and SO_4^{2-} concentrations. HCl may be transported from lower to higher layers followed by a gas-to-particle conversion.

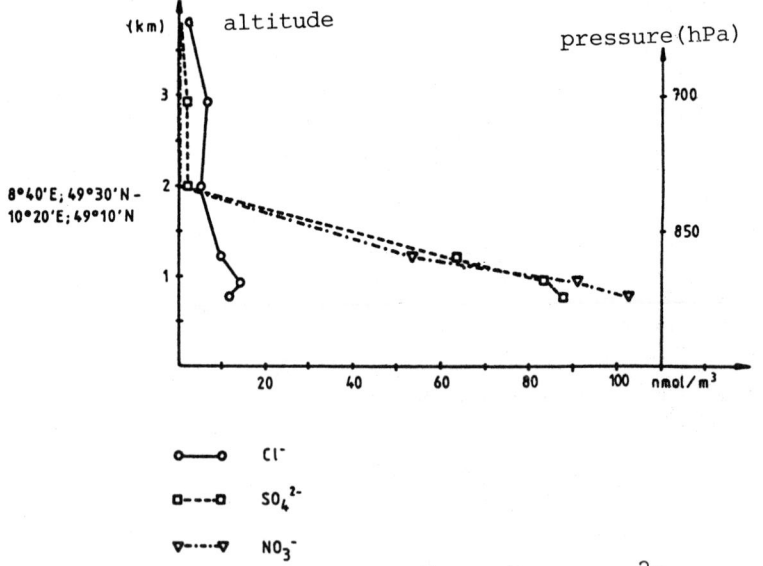

Figure 3: Vertical profile of Cl^-, NO_3^- and SO_4^{2-} (15.3.84) (10)

4. Size distribution of Cl^-, NO_3^- and SO_4^{2-}

The different size distributions of Cl^-, NO_3^- and SO_4^{2-} measured by a particle collection with an impactor are presented in Fig. 4. These distributions were measured in Darmstadt and do not represent a mean value. For a complete interpretation, the different size distributions in different altitudes are needed.

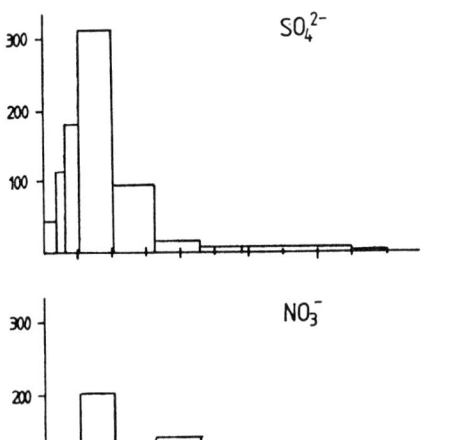

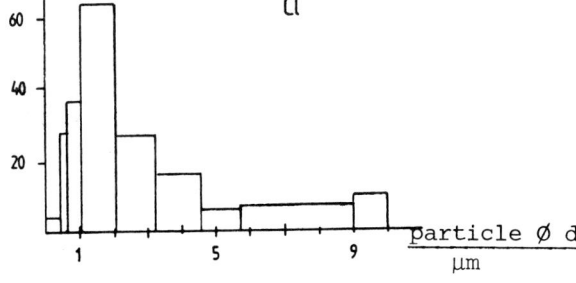

Figure 4: Particle collection with an impactor (10)

5. Diurnal variations of HCl and the anion content of aerosols

The formation of gaseous HCl is dependent on the H^+ and the NH_4^+ concentrations in aerosols and the relative humidity. Fig. 5 shows the diurnal variations of H^+, SO_4^{2-} and NH_4^+ concentrations measured in a forest near Darmstadt.

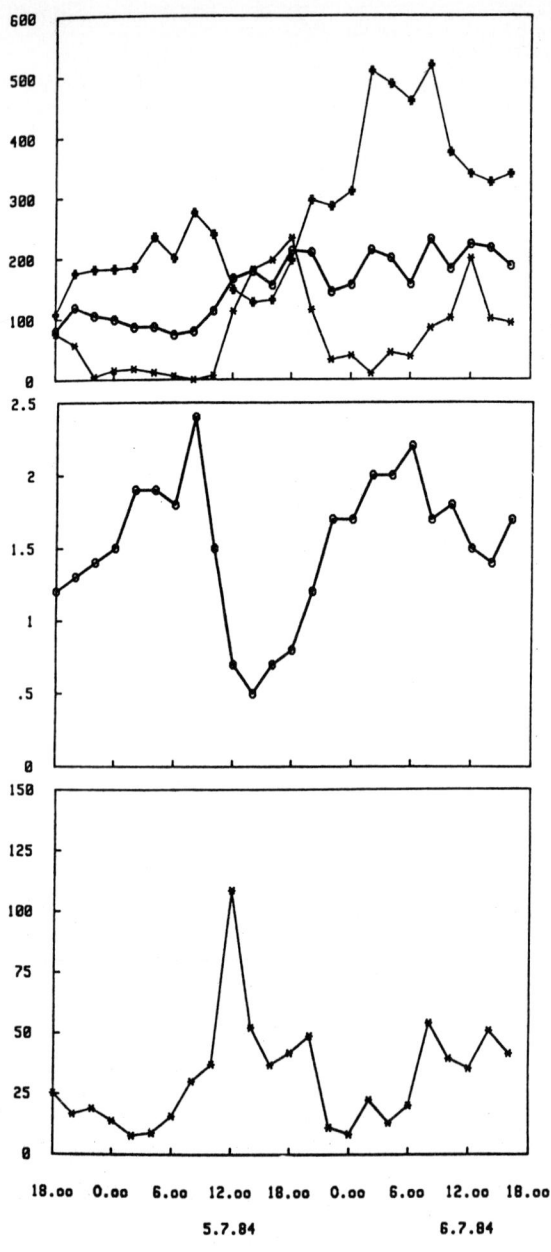

Fig. 5: Diurnal variation
SO_4^{2-} (O)
NH_4^+ (#)
H^+ (*)
(in n mol m^{-3})

Fig. 6: Diurnal variation
ratio:
$$\frac{(ammonium)-(nitrate)}{(sulfate)}$$

Fig. 7: Diurnal variation of HCl concentration (n mol m^{-3}) 5./6.7.1984

time
date

H_2SO_4 and HNO_3 may be neutralized by NH_3, so that the ratio HN_4^+/SO_4^{2-} is important. In Fig. 6, the ratio

$$\frac{(NH_4^+) - (NO_3^-)}{(SO_4^{2-})}$$

is plotted versus time. The NH_4^+ concentration is corrected by substracting the amount of NO_3^- ions present.

The ratio is 2 when $(NH_4)_2SO_4$ is present, the ratio is 1 when NH_4HSO_4 occurs. If the ratio is less than 1, additional free acid (H_2SO_4) must be present.

When the ratio is less than 1, gaseous HCl is released which is shown in Fig. 7.

6. Influence of the relative humidity on the HCl concentration

Fig. 8 presents the influence of the relative humidity on the HCl concentration. The upper line is the urban values, the lower line the rural value. The middle line corresponds to all measured values. There is a maximum at about 50 % relative humidity. At higher humidities, the HCl concentration decreases because of the solubility in the liquid surface layer of aerosols, the decrease at relative humidities less than 50 % may be due to the lower oxidation rates of SO_2 to H_2SO_4.

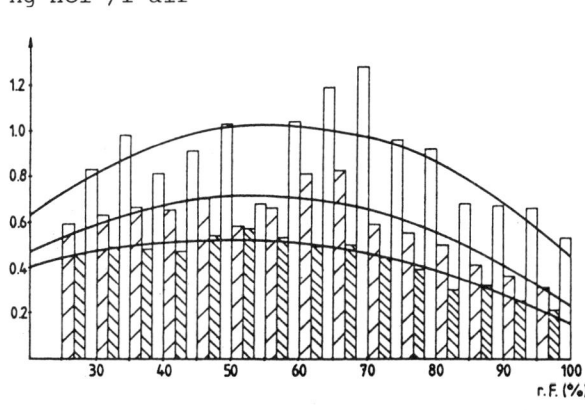

Figure 8: Influence of relative humidity on HCl concentration (9)

7. Phosgene and Cl radicals

Besides HCl, phosgene occurs in the photochemical degradation of chlorinated hydrocarbons. Fig. 9 presents preliminary results of phosgene

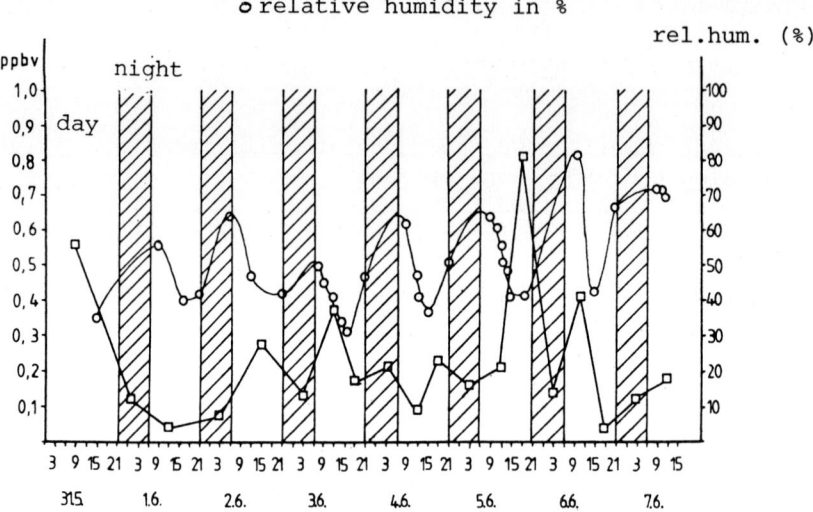

Figure 9: Phosgene measurements (12)

measurements in Darmstadt. Daily maxima and minima in the nights can be realized. The formation of phosgene was measured in smog chamber experiments in the photochemical degradation of trichloroethylene.

The tropospheric lifetimes of chlorinated hydrocarbons must be long enough to be transported to remote regions and short enough to be decomposed in the troposphere. In Fig. 10, the amount of chlorine decomposed in the troposhere is plotted versus the tropospheric lifetimes for several hydrocarbons. For a degradation in the troposphere chlorinated hydrocarbons up to CH_3Cl must be considered.

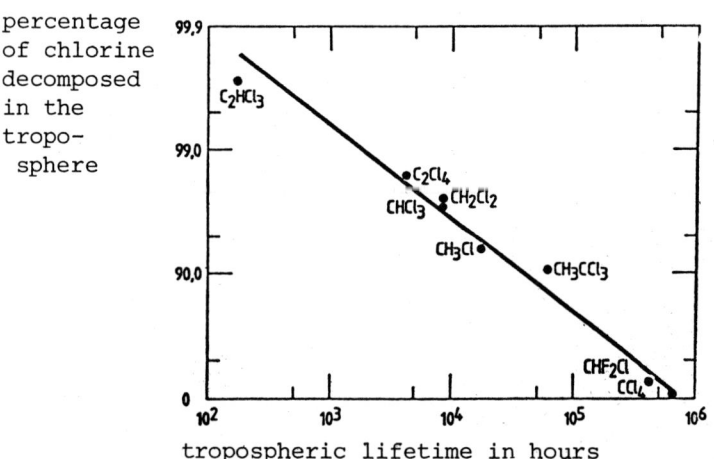

Figure 10: Chlorine decomposed versus lifetimes (13)

Tab. 4 presents the results of trichloro- and tetrachloroethylene measurements. In forest areas, a concentration of about 1 µg m^{-3} trichloroethylene occurs.

Tab. 4: Concentrations of C_2Cl_4 and C_2HCl_3 in urban and forest areas (14,15)

µg/m³	Darmstadt urban	Darmstadt forest	Ulm urban	Ulm forest
C_2Cl_4	7.7	0.50	2.03	0.35
C_2HCl_3	9.1	1.10	6.20	0.80

The main source for Cl radicals is the reaction of HCl with OH radicals, sinks are the reactions of Cl with hydrocarbons. With usual concentrations of HCl, OH and important hydrocarbons, a Cl radical concentration between 2×10^2 and 3×10^3 cm^{-3} is estimated.

References

1. Frank, H. EPA Newsletter, in press

2. Matusca P., Schwarz B., Bächmann K., Atmos. Env. 18, 1667 (1984)

3. Bächmann K., Proceedings of the 2nd Coll. Anal. Chemistry, Duisburg

4. Vierkorn-Rudolph B., Bächmann K., Schwarz B., Meixner F.X., J Atmos Chem. 2, 47 (1984)

5. Fuchs G., Lisson E., Schwarz B., Bächmann K., Fres. Z Anal.Chem. 320 498 (1985)

6. Vierkorn-Rudolph B., Rudolph J., Meixner J.X, Roth E.P., Bächmann K. Proceedings of the 3rd Europ. Symposium on the Physico-Chemical Behaviour of Atmospheric Pollutants, Varese, Italy (1984)

7. Behnke W., Koch W., Nolting F., Zetzsch C., 2nd European Symposium on Chemistry and the Environment, Lindau (1983)

8. Schenk O., EPA Newsletter, April 1985

9. Schwarz B., Thesis TH Darmstadt (1985)

10. Fuchs G., Lisson E., Bächmann K., unpublished results

11. Müller J. 4th Workshop Ion Chromatography, Dionex, Wiesbaden 1985

12. Reineke F., Bächmann K., unpublished results

13. Derwent R.G., Eggleton A.E.I., Atmos.Env. 12, 126 (1978)

14. Tille J., Bächmann K. unpublished results

15. Class Th., Mayer P., Ballschmiter K. 2nd European Conference on Chemistry and the Environment, Lindau (1984)

Effects on Ecosystems

WATERSOLUBILITY OF HEAVY METALS IN DEPOSITION SAMPLES -INTERPRETATION AND PREDICTION OF BIOAVAILABILITY

E. Rohbock
Battelle Institut e.V.
Am Römerhof 35
D-6000 Frankfurt am Main
West Germany

ABSTRACT. The measured soluble and insoluble fractions of heavy metals in precipitation water and dry deposition samples are interpretated based on the theoretical equilibrium constants and the major composition of the ionic composition of the aqueous solutions. The results are that the soluble fractions of lead, cadmium and manganese·consist of bivalent ions. Soluble chloro-, nitrate- and sulfo-complexes and hydrolised species are neglectible for the mean chemical composition in deposition samples. The soluble fraction of iron is a complex mixture of various hydrolised species, trivalent ions and sulfocomplexes. The dominating species for the pH range of pH 4 to pH 6 are the $Fe(OH)_2^+$-complexes.
Based on the knowledge of the composition of soluble fractions the chemical composition of the insoluble fractions can be estimated by the solubility products. Insoluble compounds are existing for manganese and iron in form of oxides and hydroxides. The measured insoluble lead and cadmium compounds in reality are adsorbed ions onto insoluble matter. The adsorption capacities of the adsorption onto manganese and iron-oxids are calculated. As the adsorbed toxic lead and cadmium ions are easily removable by a decrease of pH the determination of the soluble form of toxic metal is not sufficient for an assessment of the environmental hazards. The potential bioavailable compounds are gathered by an acidification to pH 3 prior to filtration of the samples.

1. INTRODUCTION

Airborne metals are removed from the atmosphere by wet and dry deposition and deposited into terrestian ecosystems. Toxic metals emitted into the atmosphere by anthropogeneous sources contaminate plant surfaces, soils and natural waters. For the assessment of the environmental risk it is necessary to know the chemical composition of the metals, the watersolubility and the bioavailability. It is not sufficient to measure the total amount of deposited metals. The insoluble chemical compounds of even a toxic metal can be regarded as inert compounds, whereas the watersoluble fraction represents the hazardous and toxic compounds which are bioavailable.

It is a main desire in large deposition monitoring networks to detect a large spectrum of elements and compounds. In many cases only one sample is available so that this desire confrontates with the demand for accuracy for single components. In many cases the chemical analyses of elements require different pretreatment procedures which are not compatible for other elements. Up to the present date a lot of different sampling procedures are existing. Some investigations are based on the analysis of bulk samples (dry and wet deposition), other analyse dry and wet deposition separately. In some networks the only fraction which is analysed in deposition samples is the soluble fraction. In other networks the total amount (soluble and insoluble fractions) is analysed together. A comparison of the data gained by the different methods is not possible without having a profound knowledge of the chemical and physical forms of the metals in deposition samples.

The object of the present investigation is to interpretate the data of metal solubility in deposition samples which have been collected in a network in western Germany over a period of 2 full years.

2. SOLUBILITY OF METALS IN AQUEOUS SYSTEMS

Metals in aqueous solutions are existent in various chemical and physical forms which are summarized below:

Form	Example
1. free metal cations	Pb^{2+}
2. anorganic complexes	$CdCl^{+}$
3. Organo-metal compounds	$(CH_3)_4Pb$
4. Organic complexes, Chelates	
5. Metal species bound on highmolecular organic material Metal lipids	
6. Metal species in form of dispersed colloids	$FeOOH$
7. metals adsorbed on colloids	$FeOOH$
8. Large particles of insoluble material	(Silicates)

The species of 1 to 4 give the soluble fraction /1/, the species of 5 to 8 give the insoluble fraction bound on dispersed material. As the diameters of the different forms vary from the submicrometer range to diameters of 10 µm and larger a distinction of soluble and insoluble fraction can be gained by filtration. The use of membranefilters (0.45 µm poresize) has become an unwritten standard in most laboratories /2/. The specific determination of free cations in the unfiltered solution has been proved by Gonzales et al. /3/.

The distribution of the metals in the different forms is determined by various parameters. The main parameters are:

* pH- value
* Redox conditions
* Temperature
* Concentration of complexing agents
* Particle size
* Time of dissolution
* Concentration of adsorbing material
* Mass concentration of the metals in the solution

Most of the parameters vary within a certain range for deposition samples. For example the pH-values of the samples are in the range of pH 6 to pH 4 (2 orders of magnitude). The same variance is found for other components. As the result the distribution of soluble/insoluble fractions in deposition samples are highly variable.

To enable an interpretation of the solubility the parameters have to be kept to fixed values as far as possible. The solution time and temperatures are kept to standard conditions. The other parameters have to be measured or have to be assumed.

3. SAMPLING AND CHEMICAL ANALYSIS

3.1 Sampling Procedure

Dry and wet depositions of airborne metals are sampled separately by aid of an automatic deposition collector. A detailed description of the instrument and the sampling characteristics are given elsewhere /4/.

Wet deposition is collected on daily basis in polyethylene (PE) funnels and 1 l PE bottles. The dry deposition is collected in highwalled glas vessels. The sequence of sampling for dry deposition is 14 days in order to receive a mass sufficient for the analysis.

The deposition samples have been collected at different stations which are distributed over the whole area of the Federal Republic of Germany. A detailed characteristic of the stations is given by Georgii et al. /5/.

3.2 Chemical analysis

The distinction of watersoluble and waterinsoluble fraction of precipitation water is performed by filtration through membrane filters (poresize 0.45 µm). The samples of dry deposition (14-day samples) are dissolved in 30 ml deionized water by aid of an ultrasonic bath. The filtration was done after a standtime of 24 hours in order to get equilibrium conditions. This period has been proved to be sufficient for the complete dissolution of atmospheric particles /6/.
The scheme of the sample pretreatment and the chemical analysis is given elsewhere /7/. After filtration an aliquot of the filtrate is analysed for the major anions by ionchromatography. The soluble and insoluble

fractions are analysed after chemical treatment by flameless AAS for lead, cadmium, manganese and iron.

The amount of 30 ml water for dissolving the dry deposited material proved to give concentrations for most metals in the same range as the concentrations found in precipitation water (Table 1). The variances found for manganese and iron result mainly from the fact that dry deposition of these metals is the dominant sink.

Table 1 Average concentrations of metals and major ions in precipitation water and dry deposition solutions (Mol/l)

	Precipitation water	Dry deposition 1)
Lead	$1. \cdot 10^{-7}$	$1.6 \cdot 10^{-7}$
Manganese	$1.5 \cdot 10^{-7}$	$10. \cdot 10^{-7}$
Iron	$1. \cdot 10^{-7}$	$19. \cdot 10^{-7}$
Cadmium	$0.7 \cdot 10^{-8}$	$1. \cdot 10^{-8}$
Sulfate	$5. \cdot 10^{-5}$	$10. \cdot 10^{-5}$
Nitrate	$5. \cdot 10^{-5}$	$7. \cdot 10^{-5}$
Chloride	$3. \cdot 10^{-5}$	$6. \cdot 10^{-5}$
pH-value	4.2-4.8	5.7

1) Deposition-sample of 14 days dissolved in 30 ml deionized water

4. SOLUBILITY OF METALS IN DEPOSITION SAMPLES

The average values of the watersoluble fractions in precipitation water and dry deposition solution are given in figure 1. The figure makes obvious that the watersoluble/insoluble distributions of precipitation water and dry deposition solution are quite different. In precipitation water the soluble fractions of lead, cadmium and manganese amount to more than 80%, whereas the soluble fraction of iron has an average value of 60%.

In the solution of dry deposition the metals are mainly in insoluble form. For manganese and cadmium the soluble fraction amounts to about 50%, for lead an average value of 20% is found and for iron the soluble fraction is in the range of 10%. Large regional varabilities are found for the soluble/insoluble distribution of iron in deposition samples. Though local and regional differences of the solubility are found a systematic trend is not obvious.

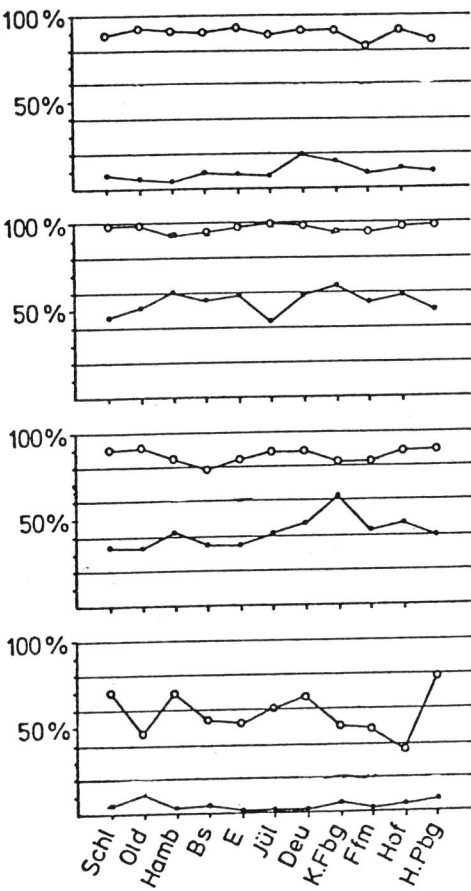

Figure 1 Average values of the relative solubility of lead, cadmium, manganese and iron in precipitation water and solution of dry deposition at various sampling sites in western Germany.

5. COMPOSITION OF THE SOLUBLE FRACTION

5.1 Basic assumptions

The interpretation of the chemical compounds in the soluble fraction is based on measured and assumed parameters. The measured parameters of the solutions are:
-- pH value
-- Sulfate concentration
-- Nitrate concentration
-- Chloride concentration
-- Soluble metalconcentration

Assumptions have been performed for following parameters:

-- Redox conditions
-- CO_2-concentration in solution
-- Temperature
-- Concentrations of halogenids not been measured
-- Concentration of ammonia in solutions
-- Saturation of atmospheric gases

The redox limit of the aqueous systems is given by the reaction

$$H^+ + e^- + 1/4\ O_2(g) \rightleftharpoons 1/2\ H_2O \qquad (1)$$

The value of the equilibrium expression K^0 is $10^{20.78}$. As the activity of water is very near unity the equilibrium expression in logarithmic form becomes

$$pe + pH = 20.78 - 1/4 \log O_2(g) \qquad (2)$$

The precipitation water is a priori in equilibrium with the atmospheric gas composition. Equal conditions can be assumed for the solutions of the dissolved dry deposition after the standtime of 24 hours. For both solutions a (pe + pH) - value of 20.61 is considered which is obtained from equation 2 by substituting $O_2(g)$ by 0.20 atm.
For the same reason (assumption of equilibrium to atmospheric gas-concentrations) the pCO_2-value is assumed to 3.43 which represents a CO_2 - partial pressure of $330. * 10^{-6}$ atm.

Concentration values of other halogene-compounds in deposition samples are not available, but it can be assumed that in average conditions (not in the vicinity of specific sources) the bromide and jodid concentration will not exceed the chlorid values.

Maximal ammonia concentrations are assumed in the range of the SO_4^{2-}-concentrations.

5.2 Metal species in the soluble fraction

In the soluble fraction the metals are existent in ionic and in form of anorganic complexes. Organic compounds of the metals being investigated in this paper are existent only for lead in form of ethyl-methyl-lead compounds (antiknock agents). The portion of these compounds to total atmospheric concentrations is less than 4% /8/. As these organic compounds are less watersoluble the amount of organic lead compounds in the deposition samples is neglectible.

Based on the equilibrium constants summarized by Lindsay /9/ the relative distribution of the various anorganic compounds is determined.

Lead and cadmium are stable in aqueous solutions only in the bivalent form, whereas for manganese and iron the bivalent and trivalent form are possible. The dominating form is determined by the redox conditions of the system. Table 2 shows the relations of the bivalent to trivalent concentrations for two pH-values (pH 4 and pH 6). Manganese is existing in the bivalent form, whereas iron exists in the trivalent form.

Table 2 Distribution of the bi- and trivalent species of manganese and iron in deposition samples

Reaction	log K°	log (Me^{2+}/Me^{3+}) pH 4	pH 6	dominating Oxidation value
$Mn^{3+} + e^- \rightleftharpoons Mn^{2+}$	25.55	8.94	10.94	Mn^{2+}
$Fe^{3+} + e^- \rightleftharpoons Fe^{2+}$	13.04	-3.57	-1.57	Fe^{3+}

Hydrolised complexes are of minor importance for manganese, lead and cadmium. The relation of the hydrolised species of lead to free cations described by the ratio $\log(PbOH^-/Pb^{2+})$ is in the range of -3.70 at pH 4 and -1.70 at pH 6. For manganese and cadmium the hydrolised species are of even lower importance.

For iron the hydrolised form is the dominating form. The distribution of the different species is determined by the pH-values as shown in figure 2. In the range between pH 4 and pH 6 which is the most relevant range for precipitation water the dominating form is the $Fe(OH)_2^+$ - ion. Only for extremely acid conditions (pH-value lower than pH 2) the Fe^{3+} - ion is the dominating form.

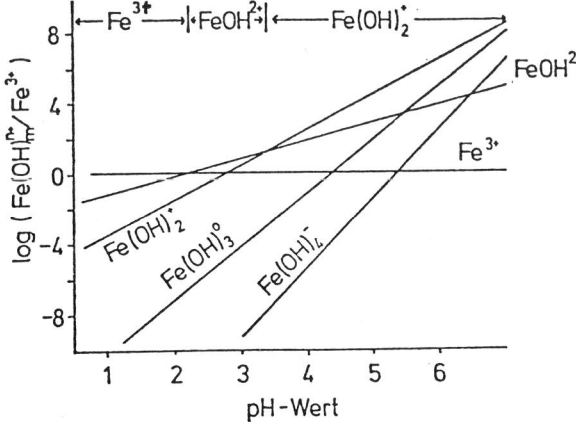

Figure 2 Distribution of hydrolised and ionic species of iron in aqueous solutions representing deposition samples

The importance of anorganic complexes can be summarized in the
following way:

- Nitrate and chloride complexes are of little importance for the metals
  being considered. The same is true for other halogenide-complexes.

- Sulfate-complexes are found for iron and for lead in the presence of
  increased sulfate-concentrations.
  At a mean sulfate concentration of $5.\times 10^{-5}$ Mol/l the relation of
  $FeSO_4^+/Fe^{3+}$ is in the range of 0.71. which means that the concentrations of the soluble sulfate complex are in the range of the free
  Fe^{3+}-ions.
  For lead the sulfate complex is important only for increased sulfate-concentrations. At a sulfate-concentration of 10^{-4} Mol/l this complex
  accounts to 4% of the total soluble form. At sulfate-concentrations
  of 10^{-3} about 30% of the soluble form consists of the sulfate-complex.

- Carbonate complexes are of minor importance.

- The same is true for ammonia-complexes under the assumption that the
  ammonia concentrations are lower than 10^{-4} Mol/l.

6. SPECIFICATION OF THE UNSOLUBLE FRACTION

Based on the knowledge of the distribution of metal species in soluble
form and based on the knowledge of the solubility constants of the
major chemical metalcompounds the composition of the insoluble fraction
is assessed. The assessment is based on the fact that the 'insoluble'
compounds will be dissolved until the solubility equilibrium (saturation
concentration in soluble form) is reached.

The saturation concentrations of the main compounds in relation to pH
value and concentration of complexing agents are given in the figures 3
to figure 6. The soluble concentrations for lead, manganese and cadmium
in molare units are assumed as the bivalent ionic form. The soluble iron
concentration results from the complex mixture of Fe^{3+}-, $FeOH^{2+}$-, $Fe(OH)_2^+$-
, $Fe(OH)_4^-$ and $FeSO_4^+$-ions.

The measured lead and cadmium concentrations in precipitation water are
found 2 to 4 magnitudes below the saturation concentrations of the most
insoluble compounds. This means that the most insoluble compounds (lead-
,cadmium-sulfates and oxides) are dissolved completely. The conclusion
is that the measured insoluble fractions of lead and cadmium in reality
are ions adsorbed onto insoluble substances (see Chapter 7).

For manganese and iron insoluble compounds are existent in form of oxides
and hydroxides. The saturation concentrations of these compounds are in
the range of concentrations which are measured in the solutions.

Only for few cases the soluble iron and manganese concentrations are below the saturation concentrations of FeOOH, Fe_2O_3 and MnO_2. For these cases total amount of iron and manganese compounds are dissolved. The insoluble fraction is not found. For the other cases when the soluble concentrations exceed the saturation concentration the insoluble fraction consists of the specified compound. This is discussed in detail for one example:
At a given pH-value of pH 4 and a measured soluble Mn-concentration of 10^{-6} Mol/l the insoluble fraction consists of MnO_2. Other compounds will be dissolved and are not stable in the insoluble form. At pH-values of pH 5.7 and at manganese concentrations of 10^{-6} Mol/l (mean values of dry deposition solution) the manganese compounds MnOOH and Mn_2O_3 are existent in the insoluble form additionally.

The existence of insoluble compounds counteracts to a further acidification of the samples. As the dissolution is H^+-ion consuming the decrease of pH-values will prelaminary result in a further dissolution. The dissolution of 1 Mol MnO_2 results in a consumption of 4 Mols of H^+-ions (Equation 3). For the dissolution of Fe_2O_3 to $Fe(OH)_2^+$ a netto consumption of 2 Mols H^+ per Mol is found (equation 4+6). The dissolution of FeOOH consumes 1 Mol H^+ per Mol (equation 5+6).

$$MnO_2 + 4H^+ + 2e^- \rightleftharpoons Mn^{2+} + 2H_2O \quad (3)$$

$$\tfrac{1}{2} Fe_2O_3 + 3H^+ \rightleftharpoons Fe^{3+} + \tfrac{3}{2}H_2O \quad (4)$$

$$FeOOH + 3H^+ \rightleftharpoons Fe^{3+} + 2H_2O \quad (5)$$

$$Fe^{3+} + H_2O \rightleftharpoons Fe(OH)_2^+ + 2H^+ \quad (6)$$

The results of these considerations can be summarized as follows:
As long as insoluble metals are existing the further increase in acidity is limited. The measured peak values of acidity (pH 3 and lower) have to be seen in context to missing insoluble material. As this material is incorporated into precipitation water mainly by washout (incorporation of falling droplets in the boundary layer), low pH-values will occure when low concentrations of these natural emitted metals are found. This will be true for winter season when the country is snow-covered and the release of ground material is reduced. On the other hand the pH-values will be increased (lower H^+-concentrations) when the release of insoluble metal compounds is increased. This has been proofed for a sampling site which is influenced by a nearby situated scrap yard /5/. Especially in springtime the amounts of insoluble ironcompounds are increased. In parallel an increase of the pH-values in precipitation water is obvious.

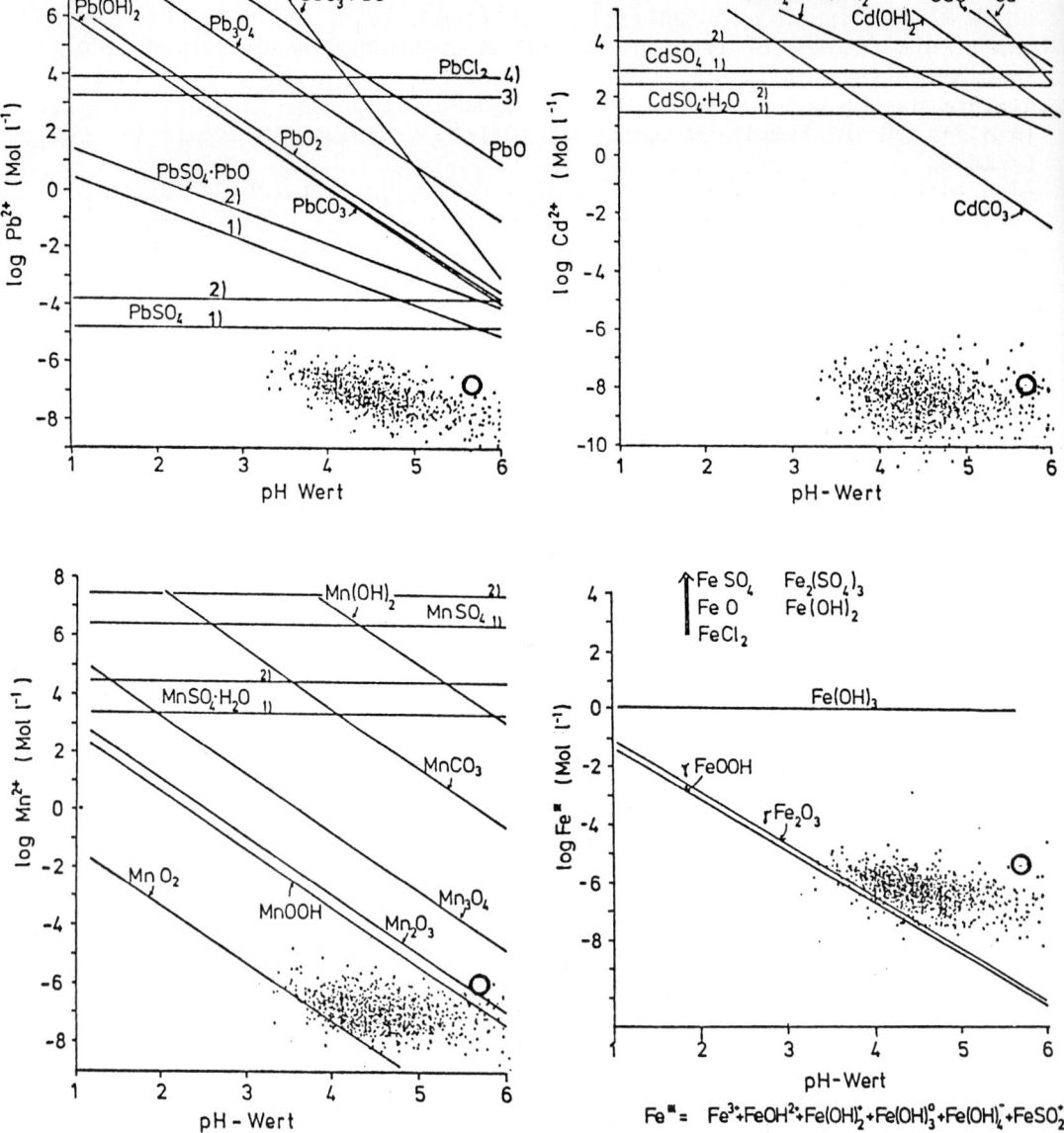

Figures 3-6 Solubility of lead, cadmium, manganese and iron compounds in aqueous solutions
Temperature 25°C   pH + pe = 20.61 $p(CO_2)g = 3.43$
1) $p(SO_4^{2-}) = 3$ 2) $p(SO_4^{2-}) = 4$
3) $p(Cl^-) = 4$ 4) $p(Cl^-) = 5$

The measured soluble metal concentrations in precipitation water are given for comparison to the saturation concentrations. The circles represent average values of dry deposition solutions

7. ADSORPTION OF LEAD AND CADMIUM ONTO INSOLUBLE MATERIAL

For lead and cadmium the existence of insoluble compounds which are found in precipitation water and solutions of dry deposition can not be explained by the existence of insoluble compounds. Even the most insoluble lead and cadmium compounds would be dissolved in the concentration ranges found in the solutions. For that reason it has to be expected that the measured insoluble fraction in reality are adsorbed lead and cadmium ions onto insoluble material. The adsorption of lead and cadmium ions onto amorphous iron- and manganeseoxides is a wellknown fact which has been investigated in the connexion with the mobility of heavy metals in natural waters /10,11/.

The adsorption capacity is prilaminary determined by the pH-value/11/. In contrary to the adsorption of Pb^{2+}-ions onto manganeseoxides which is constant in the pH-range of pH 6 to pH 4 , the adsorption capacity of ironoxides decrease rapidly with decreasing pH-value. No adsorption is observed for pH-values below pH 4. The adsorption of Cd^{2+}-ions onto amorphous ironoxides is restricted to the pH-range larger than pH 7. The adsorption capacity onto MnO_2 decreases smoothly beneath pH 6.

The measured variance of the lead and cadmium solubility in precipitation water and the solution of dry deposition in realation to pH- values show significant parallelities to these laboratory experiments.

A distinction of the adsorption capacity of manganese and iron in the present data set is not possible. The calculation of the adsorption capacities are based on the sum of insoluble manganese- and iron-oxides.

The distribution of the adsorbed and the free metal-ions depend on the available area of manganese and iron oxides. In a first approximation the available area can be substituted by the mass of insoluble oxides.

The comparison of the ratio of soluble (= free ions) concentration and insoluble (adsorbed ions) concentrations and the ratio of the sum of insoluble manganese (MnO_2) and iron ($FeOOH, Fe_2O_3$) to H^+-concentration shows a linear relation. The linear regressions are given in equations 7 and 8. Based on the mean concentration values in precipitation water and the dry deposition samples (10 stations in western Germany) the mean adsorption capacities the adsorption equilibrium are calculated.

$$\frac{[Pb_{insol}] \times [H^+]^{0.66}}{[Pb_{sol}] \times ([Fe_{insol}]+[Mn_{insol}])^{0.66}} = 10^{0.33} \qquad (7)$$

$$\frac{[Cd_{insol}] \times [H^+]^{0.5}}{[Cd_{sol}] \times ([Fe_{insol}]+[Mn_{insol}])^{0.5}} = 10^{-0.60} \qquad (8)$$

The empiric determined adsorption equilibrium constants of $10^{+.33}$ (lead) and $10^{-.6}$ (cadmium) indicate the about 10 times higher adsorption capacity of iron- and manganese oxides for Pb^{2+} compared to cadmium. This result is in connexion to the findings of Gadde and Laitinen /11/. The adsorption of Pb^{2+}- ions shows a larger pH-dependence (weight-factor 0.66) than Cd^{2+}-ions (weight-factor 0.50).

The local variance of the adsorption capacities is demonstated in the table 3. The relation of the adsorbed to free ions is calculated for a mean concentration of 10^{-5} mol/l of iron and manganeseoxides. The calculations are performed for two pH values (pH 4 and pH 6) as limits of the range which is relevant in this context.

Elevated adsorption capacities are found for lead and cadmium at the sampling sites of Schleswig, Jülich and in the Rhein-Main-area. By a factor 2 lower adsorption capacities are found at Essen, Deuselbach, Hof and H. Peißenberg.
The local varabilities can result from different physical structures of the amorphous oxides /12/ as well as from a variable composition of the soluble fraction by compounds not being measured /13/. Further investigations on this subject are desirable.

The strong variance of the adsorption capacities between pH 6 and pH 4 give indication that the adsorption prelaminary takes place onto ironoxides. This is in context to the about a factor of 50 higher insoluble iron concentrations compared to insoluble manganese concentrations.

Table 3 Local relations of adsorbed to free lead and cadmium ions based on mean Mn and Fe-oxide concentrations of 10^{-5} Mol/l

	LEAD		CADMIUM	
	pH 4	pH 6	pH 4	pH 6
Schleswig	0.6	15.4	0.06	1.45
Hamburg	0.4	17.7	0.09	0.28
Braunschweig	0.3	9.4	0.06	0.70
Essen	0.3	5.9	0.04	0.42
Jülich	0.2	18.1	0.01	1.86
Deuselbach	0.4	9.5	0.04	0.32
Kl. Feldberg	0.7	36.3	0.13	1.92
Frankfurt	0.6	21.5	0.08	1.22
Hof/Saale	0.4	9.6	0.05	0.85
H. Peißenberg	0.7	6.0	0.06	0.54

8. CONCLUSION

The soluble fraction of metals in deposition samples is primarily the result of the chemical composition of the solution. The measurement of the soluble fraction of metals therefore is not suitable for the determination of the deposition rate and the potential hazards. Especially when dry and wet deposition are collected as bulk samples the high amounts of insoluble components and the high pH-value cause a dominating adsorption of the toxic metals lead and cadmium.

The adsorption is reversible. By a decrease in pH the adsorbed metals are removed and consequently are potentially bioavailable. As the adsorption is restricted to the pH range above pH 3 the samples should be acidified to pH 3 prior to filtration.

An acidification to pH values of pH 2 and lower to the opinion of the author is not suitable because this would result in a dissolution of inert compounds which in reality will not be dissolved.
Nevertheless the primary acidification causes an incompatible procedure for measurements of pH and major anions. If a parallel analysis of metals and anions is desired, the insoluble fraction of the filtered samples have to be leached by the acidificied solution after take away an aliquot of the solution for analysis of pH and anions.

9. LITERATURE

/1/ Stumm, W. and Keller,L.(1984) Chemische Prozesse in der Umwelt: Die Bedeutung der Spezifizierung für die Dynamik der Metalle in Gewässer, Boden und Atmosphäre in: Metalle in der Umwelt, hrsg. E. Merian, Verlag Chemie Weinheim, 21-34.

/2/ Nürnberg, H.W., Valenta, p.,Nguyen, V.D. (1982) Wet deposition of toxic metals in the Federal Republic of Germany in: Deposition of atmospheric pollutants, hrg. H.-W. Georgii und J. Pankrath, D. Reidel Publishing Comp., Dordrecht 143-157.

/3/ Gongalves, M.L.S., Sigg, L., Stumm., W. (1985), Voltametric methods for distinguishing between dissolved and particulate metal ion concentration in precence of hydrous oxides, Environ. Sci. Techn. 19, 141-146.

/4/ Rohbock, E., Georgii, H.-W.(1985) Problems of the determination of atmospheric deposition by automatic wet/dry deposition collectors, Special Environ Report 16, WMO- No 647, 540-550.

/5/ Georgii, H.-W., Perseke,C., Rohbock, E. (1982) Feststellung der Deposition von sauren und langzeitwirksamen Luftverunreinigungen aus Belastungsgebieten, Report Project 10402600, UBA Berlin.

/6/ Lindberg, S.E., Harriss, R.C., (1983) Water and acid soluble
Trace metals in atmospheric particles,
J. Geophy. Res. 88 (66), 5091-5100.

/7/ Rohbock.,E, Georgii,H.-W., Perseke,C., Kins,L.(1982)
Bestimmung von Schwermetallen im Regenwasser in:
Atomspektrometrische Spurenanalytik, hrsg. B.Welz,
Verlag Chemie, Weinheim, 295-304.

/8/ Rohbock, E.,Georgii, H.-W., Müller, J. (1981) Measurements of the
gaseous lead alkyls in polluted atmospheres, Atmos. Environ.
14, 89-98.

/9/ Lindsay, W.L.(1975) Chemical equilibria in soils,
J. Wiley and sons, New York

/10/ Lion, L.W., Altmann, R.S., Leckle J.O. (1982), Trace metal adsorption
characteristics od estuarine particulate matter : evaluation of
contributions of Fe/Mn oxide and organic surface coating,
Environ. Sci. Techn. 16, 660-666.

/11/ Gadde, R.R., Laitinen, H.A. (1974) Studies of heavy metal adsorption
on iron and manganese oxides, Anal. Chem. 46 (13), 2022-2026.

/12/ Crosby, J.A. et al. (1983) Surface areas and porosities of Fe(III) and
Fe(II)-derived Oxyhydroxides, Environ. Sci. Techn. 17, 709-713.

/13/ Benjamin, M.M., Leckle, J.O., (1982) Effects of complexation by
$Cl^-, SO_4^{2-}, S_2O_3^-$ on adsorption behaviour of Cd on Oxide Surfaces,
Environ. Sci. Techn. 16, 162-170.

DRY DEPOSITION, RETENTION AND WASH-OFF PROCESSES OF HEAVY METALS IN BEECH CROWNS: ANALYSIS OF SEQUENTIALLY SAMPLED STEMFLOW

M. Kazda and G. Glatzel
Institute of Forest Ecology
Universität für Bodenkultur
Peter Jordanstr. 82
1190 Vienna, Austria

ABSTRACT: To gain better understanding of dry deposition and subsequent wash-off processes in beech crowns, stemflow was fractionally collected by means of sequential sampler in a beech stand of the Vienna Woods and analysed for heavy metals. Time series showed that after onset of stemflow concentrations are very high but drop sharply thereafter, reaching eventually a more or less constant level. Non linear regression was used to model the change of concentration during the duration of stemflow. The part of the curve above the final quasi-constant level indicates wash-off of previously dry deposited material. Highly significant positive correlations between heavy metal input and duration of dry weather before onset of stemflow show that dry depositon to the crowns contributes heavily to total input.

1. INTRODUCTON

Stemflow in beech stands can be extremely enriched with atmospheric pollutants (Mayer 1981; Glatzel et al. 1983). Damage to ground vegetation, soil acidification and heavy metal enrichment has been reported for the infitration zone of stemflow. (Glatzel et al. 1983; Glavac et al. 1985; Kazda and Glatzel 1984). The chemical properties of stemflow are not only influenced by the quality of the precipitation above crown level but also by dissolution of materials on the tree surface. To gain better understanding of these processes, stemflow was fractionally collected with a sequential sampler constructed for this purpose. Statistical analyse emphasized the importance of dry deposition processes to beech crowns.

2. MATERIAL AND METHODS

The experimental site is located in a mature beech stand on Exelberg, a hilltop in the Vienna Woods, about 10 km from the city of Vienna. Starting in 1983 precipitation was collected on event basis and analyzed for major components. There are 15 open samplers for throughfall precipitation and three stemflow collectors, one of them a sequential sampler. The sampler was programed to collect the first, second, fifth, tenth, twentieth, fiftieth etc. liter of stemflow water, to get more data on fast concentration changes at the beginning of an event.

The samples were acidified with HNO_3 and stored at 5 °C. Concentrations of Pb, Cd, Cu and Ni were determinated by flameless Atomic Spectroscopy using Lwow' platforms. Mn and Fe were determinated using flame technics.

To model the steep decrease of concentration of trace elements in the first liters of stemflow, non linear regression fitting was used. The formula

$$f(x) = A \cdot e^{Bx} + C \qquad (1),$$

where x is the amount of the stemflow, was suitable to express the change of concentration. For every element a separate regression was computed.

Because of falling function ($B < 0$) the value of the term at $x \to \infty$ could be written as:

$$\lim_{x \to \infty} (A \cdot e^{Bx} + C) = e^{-\infty} + C = C \qquad (2)$$

During a hypothetical endles precipitation event, the concentration of an element in stemflow would converge to the value C after wash-off of previously deposited material is completed.

To get the total amount of an element in the stemflow, formula (1) was integrated and devided into two integrals:

$$\int f(x)\,dx = \int (A \cdot e^{Bx} + C)\,dx = A\int e^{Bx}\,dx + C\int dx \qquad (3)$$
$$\qquad\qquad\qquad\qquad\qquad\qquad\text{Integral I} \quad \text{Integral II}$$

Integral I represents the input by wash-off processes and also by precipitation scavenging, as concentrations in rain are generally higher at the beginning of an event. Integral II is the base input, reflecting the concentration of the rainwater above canopy. The total amount of an element (Fig. 1) in the fractionated stem-flow is a sum of two definite integrals (Integral I und Integral II).

$$\text{Integral } I_i = A_i \int_0^c e^{B_i x}\,dx$$

$$= A_i \left[(e^{B_i x} \cdot B_i^{-1}) \right]_0^c$$

$$\text{Integral } I_i = A_i (e^{B_i c} - 1) \cdot B_i^{-1} \qquad (4)$$

$$\text{Integral II}_i = C_i \int_0^c dx = C_i \bigl[x\bigr]_0^c = \underline{\underline{C_i c}} \qquad (5)$$

A_i, B_i, C_i Parameters of the function for element i

c ... stemflow precipitation (mm)

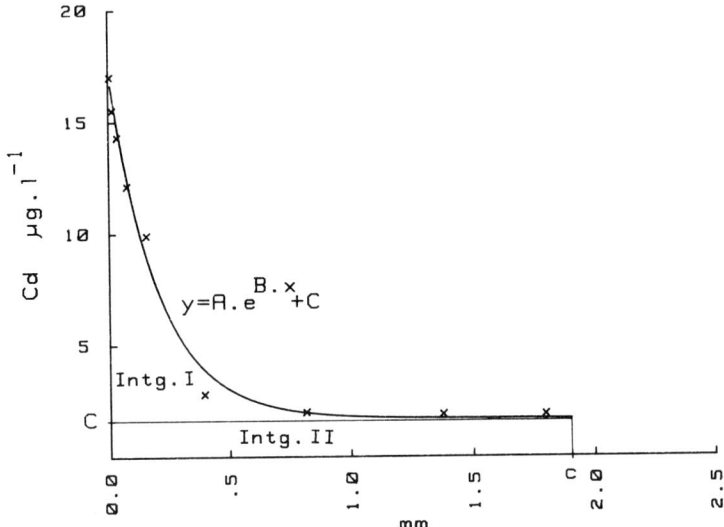

Fig. 1. Example of computed non-linear regression (Cd, event No. 3) and its integration

3. RESULTS

3.1. Dry deposition, retention and wash-off

Heavy metal input from 42 fractionated stemflow events during a 12 month period are given in the Table 1. As some events were not sampled, these figures are not the total annual input from stemflow.

Less then 20 % of the total input of lead from stemflow may be attributed to wash-off (Integral I). The reason for this, in comparison to other elements relatively low percentage, is the strong retention of lead on tree surfaces. The retention of lead in the canopy is comprehensivly documented by Mayer (1981), Höfken et al. (1981) and Kazda (1985). Integral I of the stemflow input is much larger for cadmium, copper and nickel because of higher solubility of these elements at the usually low pH of stemflow water.

TABLE 1. Heavy metal input from stemflow devided into Integral I and II and the contribution of Integral I to total input from stemflow

Element	Int I	Int II	Total stemflow input	Contribution of Int I
		$g \cdot ha^{-1}$		%
Pb	6.9	27.9	34.8	19.8
Cd	2.0	4.6	6.6	30.3
Cu	4.0	9.6	13.6	29.4
Ni	2.3	5.8	8.1	28.4
Fe	72.0	251.9	323.9	22.2
Mn	57.1	183.4	240.5	23.7

To differentiate the wash-off processes of elements studied, a multiple mean comparison test for statistical differences between the parameters of the non-linear regression (1) was computed.

Variability of the calculated initial concentrations (parameter A) of the various elements was high, however two groups could be statisticaly differentiated. Compared to copper and nickel the concentrations of cadmium are very high at the beginning of the stemflow. Manganese showed also very high values at the beginning of an event, probably due to wash-off of manganese excluded from plant tissues and not due to dry deposition.

Parameter B indicates how fast the concentration of an element decreases during a stemflow event. There are no significant differences between the elements studied. The intensity of wash-off in the order lead, nickel - copper - manganese, iron - cadmium (TABLE 2) conforms however to the general theory of heavy metal mobility. High variability shows that there are differences in wash-off velocity, probably related to the intensity of precipitation.

Parameter C represents the quasi steady state concentration after complete wash-off. Lead, copper, nickel and cadmium show no sinificant differences. The steady state concentrations of manganese are significantly lower than those of iron (TABLE 2). High variability of the cadmium-concentrations is a result of the extreme differences in deposition rates between summer and winter.

TABLE 2. Calculated parameters for concentration changes in the stemflow for various trace elements (formula (1)):
- A .. concentrations at the onset of the stemflow
- B .. intensity of concentration decrease
- C .. steady state concentration

Means underlined by a common line are not significantly different (Duncan Test, p = 0.95)

Element	Pb	Cd	Cu	Ni	Fe	Mn
A mg.l^{-1}	0.142	0.049	0.055	0.038	1.02	0.98
B	-14.2	-20.9	-16.4	-14.5	-18.9	-18.0
C mg.l^{-1}	0.040	0.017	0.020	0.012	0.45	0.29

Another approach to differentiate retention behavior of different elements is a comparision of the calculated maximum concentration at the beginning of stemflow with the steady state reached thereafter. The ratio A/C is a measure of initial to final concentration during an event (TABLE 3). It appears to be a practical parameter to judge the relative importance of dry depostion.

TABLE 3: Calculated ratio A/C
Means underlined by a common line are not significantly different (Duncan Test, p = 0.95)

Element	Fe	Cu	Pb	Mn	Ni	Cd
A/C	2.44	3.05	3.84	5.57	7.81	11.49

Iron and copper have the lowest A/C ratios. Dry deposition of these elements is appearently quite low. Cadmium has the highest A/C ratio. A high rate of dry deposition, as well as its low retention, contribute to this effect.

3.2. Factors influencing heavy metal input from stemflow

Figure 2 shows lead-input from stemflow within a 12 month period between May 1984 and April 1985 devided into Integral I and II. Remarkable peaks in November and December 1984 occured during foggy periods, when highly polluted fog-water impacted on the canopy.

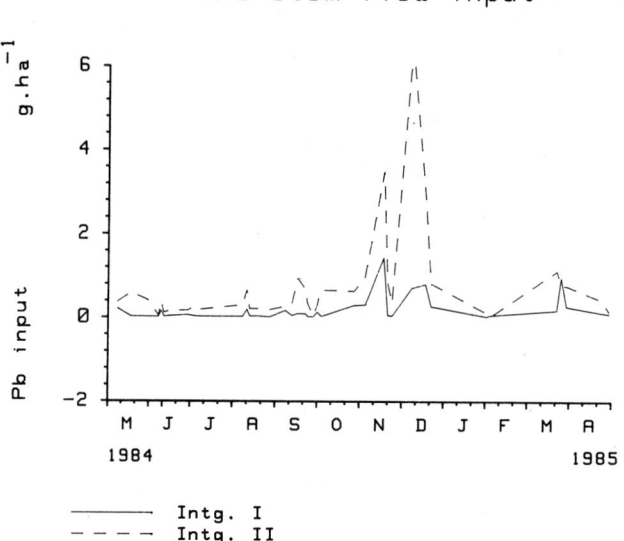

Fig. 2. Lead-input from stemflow devided into Integral I and II (see formula (3)) within a 12 month period between May 1984 and April 1985

Heavy metal deposition to beech ecosystems in the Vienna Woods shows high seasonality and is strongly influenced by meteorogical variables. Stepwise regression for heavy metal deposition with stemflow as dependent variable and meteorological parameters as independent variables showed that the variables: fog, wind direction, duration of dry weather before onset of stemflow and amount of stemflow were higly correlated with deposition rates. Highly significant positive correlations between heavy metal input and the duration of the period without precipitation prior to the onset of stemflow were calculated, which means that the input of all elements was significantly influenced by dry deposition.

For heavy metal input from wash-off (Integral I), deposition by impaction seems to be a very important mechanism. This fraction of stemflow-input was independent of the total amount of stemflow water. Stepwise regression showed that the input of lead, cadmium, nickel and manganese from wash-off (Integral I) was positively correlated with winds arriving from the Vienna metropolitan area. No statisticaly significant parameter was found to correlate copper wash-off with meteorological parameters. Compared with data from the Federal Republic of Germany there was substantially lower copper input into forest of the Vienna Woods (Kazda, 1985), probably due to unimportant copper emission in the Vienna area.

Input at steady state conditions (Integral II) was strongly influenced by temperature (increasing input at lower temperatures) and by fog. During foggy weather the input of heavy metals to the forest increased dramaticaly. More than 42 % of the annual input of cadmium occured during only 6 events with fog interception.

4. DISCUSSION

Most of the research on concentration changes during precipitation events was directed to the problem of below cloud scavenging (Coscio et al., 1982; Kins, 1982; Munger et al., 1983; Zimmermann, 1986). Generally the results show a steep drop of concentrations at the beginning of the event and then roughly constant low values. In case of convective showers the concentrations do not follows this general trend. The air masses below the clouds are not cleaned by scavenging processes because of advection of polluted air with lateral winds.

Our research on fractionated stemflow has shown that pollutant concentration in stemflow decreases rapidly regardless of the typus of precipitation event. The magnitude of the concentration drop is highly correlated to the duration of the period without precipitation before the onset of stemflow and to wind direction. Total deposition is positively correlated to winds coming from the direction of the Vienna metropolitan area.

The method of sequential collection and analysis of stemflow emphasized the importance of dry deposition between precipitation events. The contribution of high initial pollutant concentrations in the rain due to below cloud scavenging could not be accounted in this investigation. In the future a coupled sequential rain sampler instaled above the canopy will help to gain a better understanding of these processes.

AKNOWLEDGEMENT: This research was supported under grant No. 4855 by the Austrian Science Foundation.

REFERENCES

Coscio, M.R., G.C. Pratt and S.V. Krupa (1982) An automatic, refrigerated, sequential precipitation sampler. Atmospheric Environment 16, 1939-1944

Glatzel, G., E. Sonderegger, M. Kazda and H. Puxbaum (1983) Bodenveränderungen durch schadstoffangereicherte Stammablaufniederschläge in Buchenbeständen des Wienerwaldes. AFZ 26/27, 693-694

Glavac, V., H. Jochheim, H. Koenies, R. Rheinstädter and H. Schäfer (1985) Einfluß des Stammablaufwassers auf den Boden im Stammabflußbereich von Altbuchen in unterschiedlich immissionsbelasteten Gebieten. AFZ 40, 1397-1398

Höfken, K.D., H.W. Georgii and G. Gravenhorst (1981) Untersuchungen über die Deposition atmosphärischer Spurenstoffe an Buchen- und Fichtenwald. Berichte, Inst. f. Meteorologie u. Geophysik der Universität Frankfurt/Main, 46

Kazda, M. and G. Glatzel (1984) Schwermetallanreicherung und Schwermetallverfügbarkeit im Einsickerungsbereich von Stammablaufwasser in Buchenwäldern (Fagus sylvatica) des Wienerwaldes. Z. Pflanzenernaehr. Bodenk. 147, 743-752

Kazda, M. (1985) Untersuchung von Schwermetalldepositionsvorgängen aus Analysen fraktionell gesammelter Stammabflußproben und Jahresgang der Schwermetalldeposition in einem Buchenwaldökosystem des stadtnahen Wienerwaldes. Diss. Univ. Bodenkultur, Wien

Kins, L. (1982) Temporal variation of chemical composition of rainwater during individual precipitation events. In: H.-W. Georgii and J. Pankrath (Eds.): Deposition of Atmospheric Pollutants, 87-96, D. Reidel, Dordrecht

Mayer, R. (1981) Natürliche und anthropogene Komponenten des Schwermetallhaushaltes von Waldökosystemen. Göttinger Bodenkd. Ber. 70

Munger, J.W., J.M. Waldman, D.J. Jacob and M.R. Hoffmann (1983) Vertical variability and short-term temporal trends in precipitation chemistry. In: Precipitation Scavenging, Dry Deposition, and Resuspension, Eds.: H.R. Pruppacher, R.G. Semonin and W.G.N. Slinn, Elsevier, New York-Amsterdam-Oxford, 265-274.

Zimmermann, R. (1986) paper in this book

CADMIUM AND LEAD IN THE FOOD WEB OF A FOREST ECOSYSTEM

A.Fangmeier and L.Steubing
Institut für Pflanzenökologie
Heinrich-Buff-Ring 38
D-6300 Giessen

ABSTRACT. In a forested ecosystem in the Rhine-Main region, active and passive monitoring was carried out to investigate the immission load with cadmium and lead in different compartments of the system. Though the deposition during the research period was only slight, high amounts of lead were found in the organic topsoil. The dispersal of Cd and Pb in the food web was studied by analysing 19 different animal species. Cadmium showed a higher mobility in food chains than lead.

1. INTRODUCTION

The contamination of the environment with heavy metals has become an increasing problem during the last decades. Several investigations are dealing with accumulation and distribution of elements like cadmium, lead, zinc, copper, a.o. in forest ecosystems (HEINRICHS & MAYER 1980, MAYER 1981, MARTIN et al. 1982, PARKER et al. 1978, VAN HOOK et al. 1977). Further attention has been paid to the reaction of soils and forest floors on the input of heavy metals and to their distribution within the soil (ANDRESEN et al. 1980, JOHNSON et al. 1982, MILLER & MCFEE 1983, SICCAMA & SMITH 1980).
  The input of heavy metals into terrestrial ecosystems not only leads to increasing contents of these elements in soils and plants, but also increases the metal contamination of organisms at higher trophic levels, e.g. herbivorous, detritivorous and carnivorous animals and - in the end - human beings. Therefore, animals at different levels in the food web can be used as monitors of the contamination of their environment. Measurements of heavy metal contents in animals and their tissues have been carried out for several species (CHMIEL & HARRISON 1981, COUGHTREY & MARTIN 1977, HUNTER & JOHNSON 1982, IRELAND 1979a, IRELAND 1979b, KREUTZER & KRACKE 1981, MÜLLER 1982, VETTER et al. 1974, ROBERTS & JOHNSON 1978, STEUBING et al. 1984). However, the pathways of heavy metals within an ecosystem and its food web are not yet understood perfectly.

In this research, active and passive monitoring (STEUBING 1985) was carried out to investigate the immission load with cadmium and lead in an area in Western Germany. Components of each compartment of a forest ecosystem were taken as passive monitoring samples to obtain informations about the pollution at different trophic levels.

2. MATERIALS AND METHODS

The study area was located nearby the Rhine-Main-Airport Frankfurt, West-Germany. The region mainly is covered with pine (Pinus silvestris) and oak (Quercus robur) forests, growing on relative poor, sandy, and acid soils (pH ranging from 3.6 to 4.2, measured in water solution).

In the study area, four research plots were established. Sampling took place seven times in intervals of four weeks, beginning at the 6th of May 1983.

Active monitoring in the study area was carried out using standardized grass cultivars with Lolium multiflorum, which were exposed for 4, 8, 12, or 16 weeks, starting the exposure at the 6th of May 1983.

Soil samples were taken from the O_L-, O_H-, A_h-, and B_v- horizons. Six species of the natural vegetation were selected as samples for passive monitoring:

>   Pinus silvestris (needles from 1983,1982,1981, bark, and roots)
>   Quercus robur   (leaves and bark)
>   Molinia arundinacea   (leaves and roots)
>   Calamagrostis epigeios   ("   )
>   Avenella flexuosa   ("   )
>   Teucrium scorodonia   ("   )
>   Quercus robur seedlings ("   ).

Four different fungi species were harvested when the fruit bodies were ripe.

Barber traps and mouse traps were used to catch smaller vertebrates and insects (beetles, spiders, ants, etc.). Other species could be caught by hand or with a net (slugs, locusts). Earthworms were found partially by ditching out the soil and sieving it in the field or by Berlese-traps in the laboratory. Additionally, samples of two individuals of fallow-deer (Cervus dama) were obtained on the occasion of a hunting in November 1983. In the laboratory, samples of small vertebrates were dissected into different organ probes: skeleton, kidney, liver, muscle, heart, and skin or hair.

All samples (soil, vegetation and animals) were oven-dried at 105 °C for 24 hours. Vegetation and animal samples were ground in a Krups coffie-mill. Homogenisation of soil samples took place in a Fritsch ball-mill. Cadmium and lead contents were measured by ZEEMAN-AAS using an Erdmann & Grün SM 1 (STEUBING et al. 1980).

3. RESULTS

3.1. Active monitoring with Lolium multiflorum cultivars

During the research period, there was a moderate input of cadmium and lead into the research area, as monitored by analysing the exposed Lolium multiflorum cultivars. Cadmium and lead contents of the grass cultivars after different times of exposure are shown in Fig.1. According to PRINZ & SCHOLL (1975), the background levels for cadmium and lead in Lolium multiflorum amount 0.3 and 2.4 ppm, respectively. In the study area, after 16 weeks of exposure, these values were exceeded only slight for cadmium and about threefold for lead, which suggests a slight deposition of cadmium and a somewhat higher deposition of lead. The increase of cadmium and lead contents during the progress of the vegetation period could not only be observed in the exposed grass cultivars, but also in the natural vegetation (FANGMEIER et al. 1984).

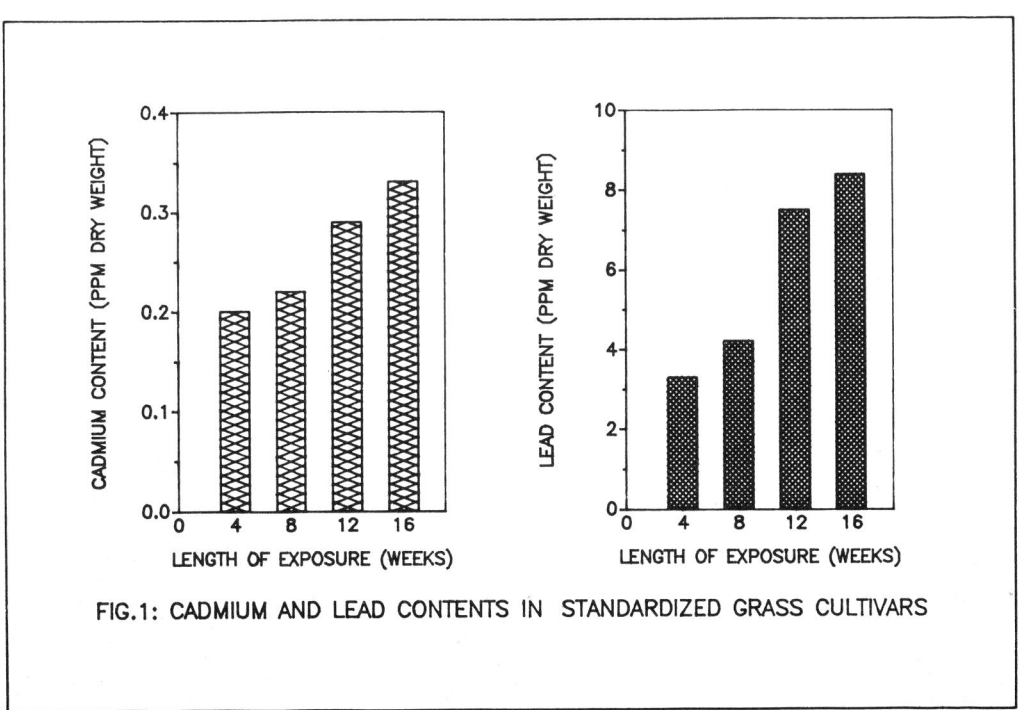

FIG.1: CADMIUM AND LEAD CONTENTS IN STANDARDIZED GRASS CULTIVARS

3.2. Natural vegetation and soil

In the following representations, the mean values of the cadmium and lead measurements during the whole vegetation period will be demonstrated.

The concentrations measured in the natural vegetation and in soil samples are shown in Fig.2 (cadmium) and Fig.3 (lead).

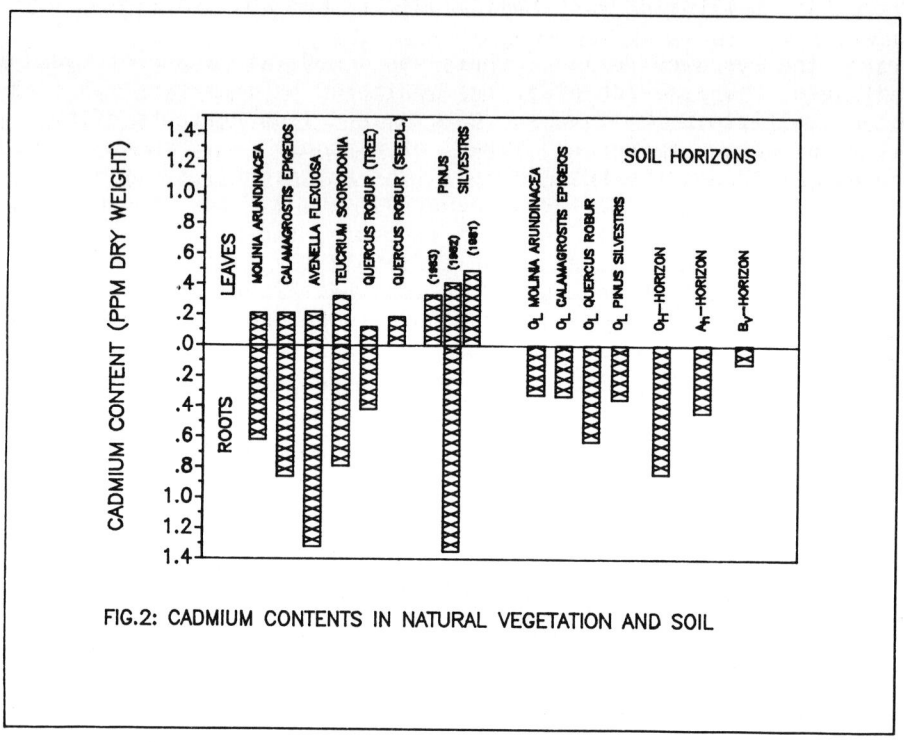

FIG.2: CADMIUM CONTENTS IN NATURAL VEGETATION AND SOIL

The cadmium contents in leaves ranged from 0.12 ppm (Quercus robur seedlings) to 0.49 ppm (in the oldest needles of Pinus silvestris). In the bark of Quercus robur and Pinus silvestris, 0.36 and 0.33 ppm Cd were measured, respectively. The highest amounts of cadmium could be found in roots (up to 1.35 ppm in P. silvestris). Litter from the O_L-horizon contained 0.32 (Molinia arundinacea) to 0.63 ppm Cd (Q.robur). Within the soil, the highest concentration was measured in the O_H-horizon (0.84 ppm Cd), followed by the A_h-horizon (0.44 ppm) and the B_v-horizon (0.12 ppm).

This situation looks very similar for lead (Fig.3). Lead contents in leaves ranged from 2.9 ppm (in the youngest needles of Pinus silvestris) to 11.3 ppm (Teucrium scorodonia). The higher amounts of lead - and cadmium - in leaves of T. scorodonia compared with the other investigated species can be explained by the relative rough leaf surface of this species, which possibly increases the deposition velocity of particulate immissions.

Bark of Quercus robur and Pinus silvestris contained quite high amounts of lead - 63.6 and 73.3 ppm, respectively. Lead levels in roots, as far as small roots with a diameter of less than 1 mm were sampled, were between 26.6 ppm (P. silvestris) and 49.3 ppm (Teucrium scorodonia). Roots of Molinia arundinacea contained only 8.1 ppm lead. This sample consisted of bigger roots with a diameter of 2 mm or more. A similar relationship of lead contents between roots of different diameters was found by VAN HOOK et al. (1977).

The lead concentrations in litter (O_L-horizon) were far above those of leaves, ranging from 32.5 ppm (Pinus silvestris) to 41.7 ppm (Quercus robur). The highest amount of lead within the plant-soil-system was found in the O_H-horizon (102.6 ppm), followed by the A_h-horizon (84.6 ppm). The B_v-horizon contained 25.7 ppm lead.

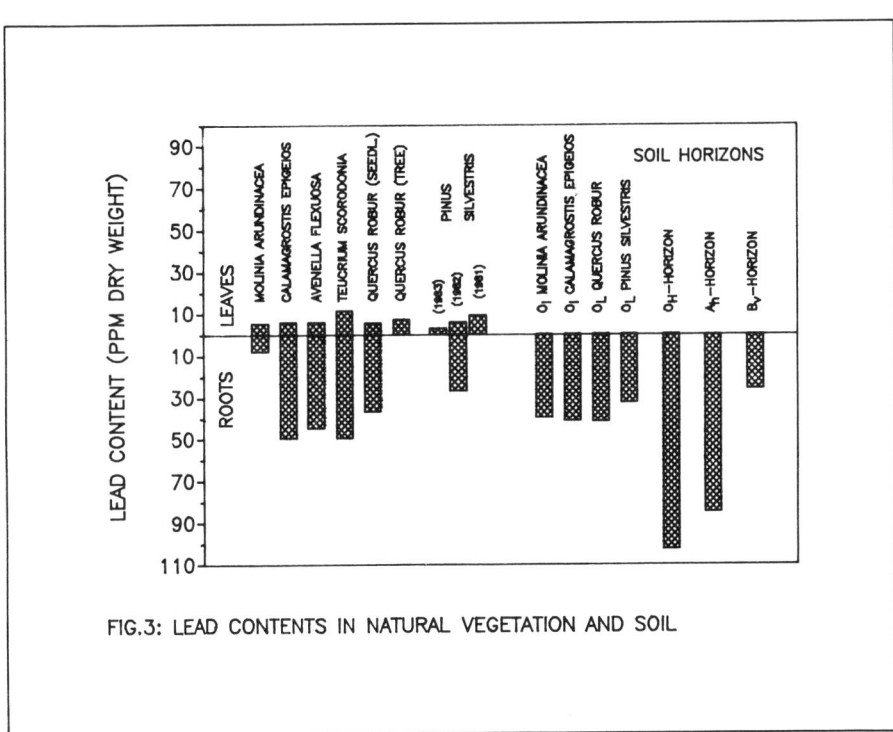

FIG.3: LEAD CONTENTS IN NATURAL VEGETATION AND SOIL

3.3. Cadmium and lead in the food web

Altogether, 19 animal species were analysed for their cadmium and lead contents. A list of these species, including the results of Cd- and Pb-measurements is given in Tab.1.

Table 1: Cd- and Pb-contents of investigated animal samples

Taxa	Species	Cd [ppm]	Pb [ppm]
Oligochaeta	Lumbricus rubellus	7.8	35.9
Gastropoda	Arion subfuscus	9.0	18.4
Arthropoda:			
Araneae	Trochosa spec.	9.6	5.6
"	Lycosa spec.	7.0	2.9
Myriapoda	Lithobius forficatus	0.86	6.0
Collembola	Tomocerus spec.	0.92	10.4
Blattodea	Ectobius silvestris	0.35	11.1
Saltatoria	Pholidoptera griseoaptera	0.19	3.2
"	Chorthippus biguttuls	0.13	1.8
"	Gomphocerus rufus	0.10	1.7
Coleoptera	Geotrupes stercorosus	0.31	6.4
Carabidae	Abax ater	0.42	2.4
"	Pterostichus melanarius	0.29	2.3
"	Pterostichus niger	0.35	2.8
"	Carabus nemoralis	0.60	2.5
Formicidae	Formica polyctena	13.8	10.5
Vertebrata	Rana dalmatina	see Fig.4	see Fig.5
	Sorex minutus	"	"
	Cervus dama	"	"

The Cd- and Pb-contents of four investigated fungi species are listed in Tab.2.

Table 2: Cadmium and lead contents in fungi

Species:		Cd [ppm]	Pb [ppm]
Macrolepiota	(cap)	12.7	11.3
mastoidea	(stalk)	5.2	6.0
Lycoperdon gemmatum		1.3	20.3
Lactarius	(cap)	13.6	3.7
chrysorheus	(stalk)	3.6	9.6
Mycena	(cap)	1.3	4.4
polygramma	(stalk)	0.59	5.8

In Fig.4 and Fig.5, the investigated samples are - as far as possible - arranged in a food web scheme.

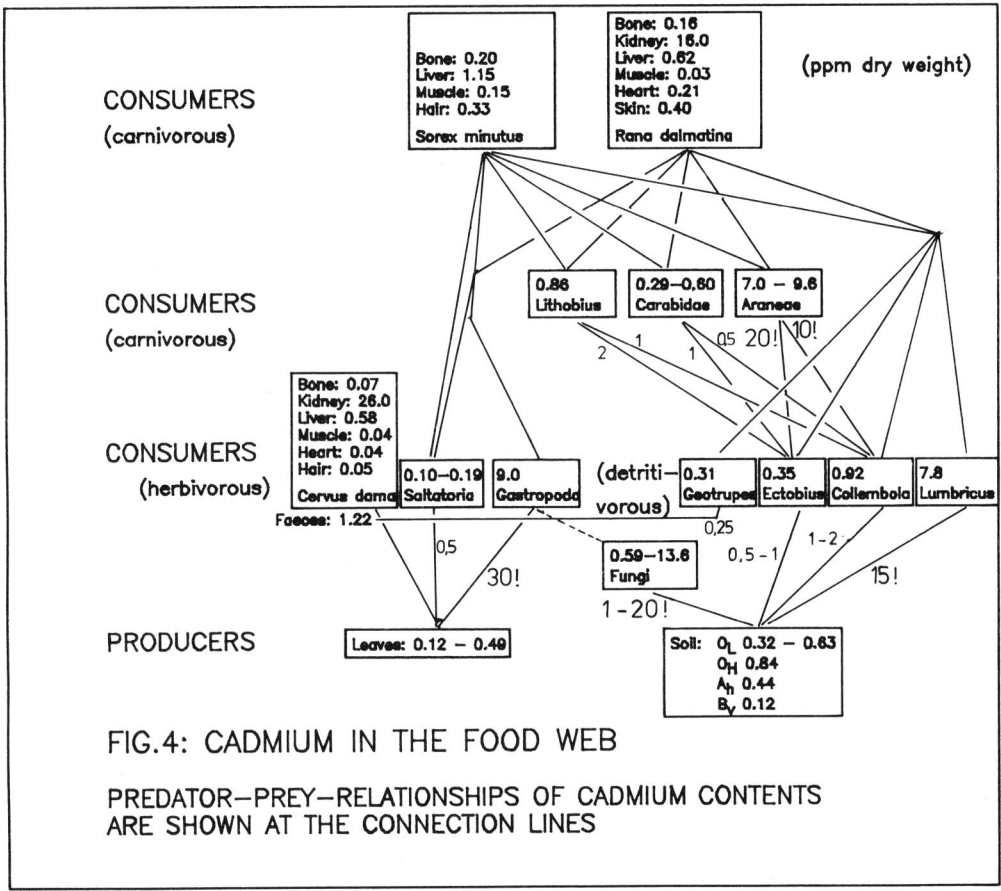

FIG.4: CADMIUM IN THE FOOD WEB

PREDATOR—PREY—RELATIONSHIPS OF CADMIUM CONTENTS ARE SHOWN AT THE CONNECTION LINES

Cadmium contents of the most of the investigated arthropod species were in the range of their estimated diet or somewhat below these values. That was true for the detritivorous species Geotrupes stercorosus, Ectobius silvestris, and the collembole Tomocerus spec., and also for the herbivorous locusts and for the carnivorous Carabidae and the millepede Lithobius forficatus. However, very high cadmium concentrations were measured in the ant Formica polyctena (13.8 ppm) and in the spiders Trochosa spec. and Lycosa spec. (9.6 and 7.0 ppm, respectively). An explanation for these high values, compared with other arthropods, must be searched in nourish-physiological qualities of Formica and Lycosa or Trochosa (food turnover rates, extra-intestinal digestion in spiders, etc.). Formica polyctena could not be arranged in the food web. Its main diet are Lachnides, which could not be analysed.

High amounts of cadmium also were measured in the slug Arion subfuscus and in the earthworm Lumbricus rubellus, both containing manifold the Cd-values of their supposed food source (30-fold for Arion and 15-fold for Lumbricus).

Within organ probes of vertebrates, the highest cadmium concentrations were found in kidneys (up to 26 ppm for Cervus dama). Also the livers were contaminated, while bones, muscles, hearts and skin or hair contained only little cadmium.

An accumulation of Cd took place in the investigated fungi. Depending on species, the cadmium contents were up to 20-fold above those measured in the O_L- and O_H-horizon.

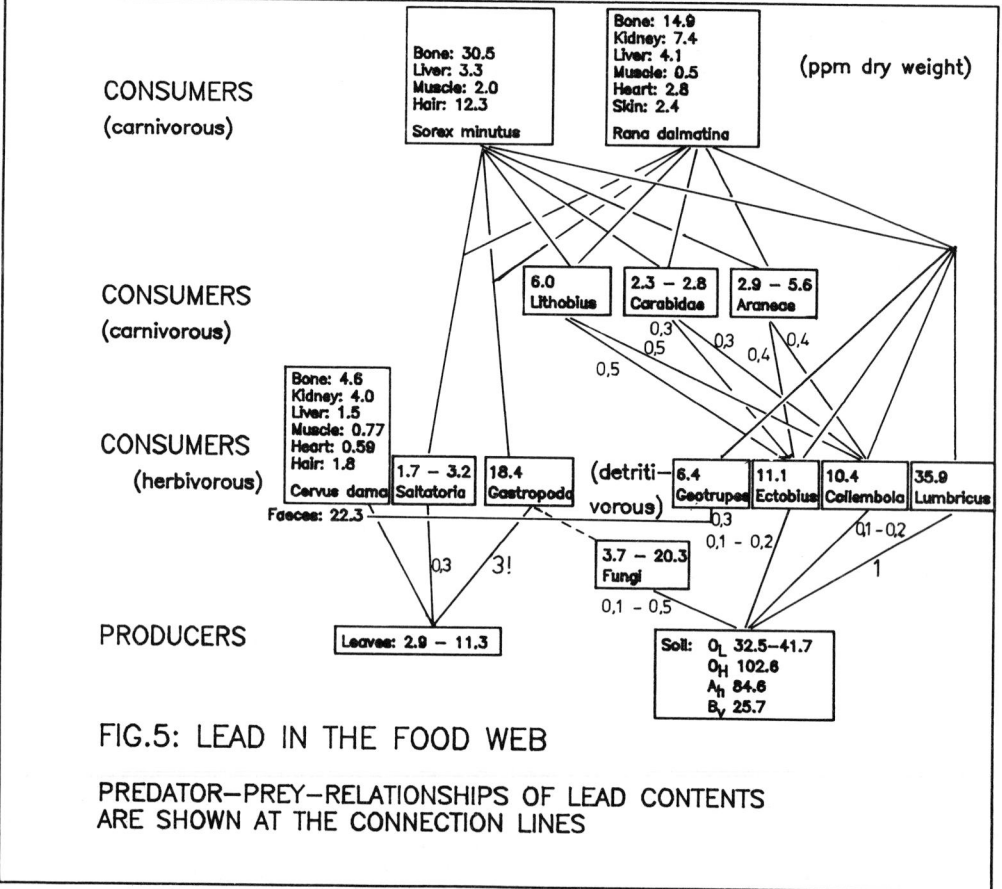

FIG.5: LEAD IN THE FOOD WEB

PREDATOR—PREY—RELATIONSHIPS OF LEAD CONTENTS ARE SHOWN AT THE CONNECTION LINES

This situation was somewhat different for lead.

Generally, lead concentrations in evertebrates were below those of their estimated diets. There were two exceptions from this "rule": Arion subfuscus contained threefold the lead amount of leaves, and the concentration in Lumbricus rubellus was in the same range as in the O_L-horizon.

In vertebrates, the highest lead contents were found in bones, followed by kidney and liver. Muscles and hearts were contaminated only slightly, the concentrations in hair or skin were above those in muscles.

In contrast to cadmium, lead was not accumulated by the investigated fungi species. The concentrations measured in fruit bodies were far below those in O_L- and O_H-horizons.

4. DISCUSSION

In the study area, active and passive monitoring was carried out to investigate the cadmium and lead contamination of the ecosystem. Both methods gave an information about the height of the pollution.

The deposition of cadmium and lead during the research period was only slight, as documented by the relative low contents in Lolium multiflorum cultivars. As far as background levels from unpolluted areas for the investigated plant and animal samples were available, the amounts of cadmium and lead measured in this study were in this range or slightly exceeded these values. However, lead concentrations up to 100 ppm could be measured in the litter layer, indicating that the forest floor must be looked at as sink for atmospheric input of heavy metals by dry and wet deposition (MAYER 1981, VAN HOOK et al. 1977, PARKER et al. 1978, SICCAMA & SMITH 1980). Of course, heavy metals are not bound irreversibly in the litter layer, but can partially be leached from the soil or be taken up by plants with the transpiration water or by detritivorous organisms. The strongness of the binding of heavy metals to the organic topsoil is different for each element (MILLER & MCFEE 1983), so that the soil contains very different percentages of the total pool of each individual element. So, HEINRICHS & MAYER (1980) found that up to 27 % of total cadmium within an ecosystem were contained in the biomass, while the lead content of the biomass only came up to 1 % of the total amount.

The different behavior of cadmium compared with lead within the ecosystem could also be observed in this study. Cadmium appears to be more mobile than lead, with a higher accumulation of Pb than of Cd in the organic topsoil on the other hand. This was indicated by the relation of element contents in leaves, O_L- and O_H-horizon. This ratio was 1 : 1.5 : 3 (leaves : O_L-horizon : O_H-horizon) for cadmium and 1 : 6 : 16 for lead, meaning a far higher accumulation of lead in the topsoil than of cadmium.

Within a plant, cadmium generally is transported more easily than lead. The greatest portion of lead applied to a plant via the soil is retained in the fine roots, the endodermis building a barrier for further transport (KELLER & ZUBER 1970). This is not true for cadmium (TURNER 1973). So, in this study, the amount of lead in investigated fine roots with a diameter of one mm or less was about 5-fold of that found in bigger roots with a diameter of two mm or more. In opposite, cadmium contents were very similar both in fine roots and bigger roots, indicating a transport of cadmium from fine roots to the plant tops.

Another indicator for the different mobility of Cd and Pb within the plant was the ratio fine-root-content : leaf-content. This ratio amounted about 6 : 1 for lead and about 3 : 1 for cadmium. This ratio would have been even higher for lead, if the lead deposition in the study area had been as low as the cadmium deposition, causing lower Pb-contents in leaves as measured under the factual conditions.

In the food web, the distributions of cadmium and lead were quite unequal. Generally, with one exception (Arion subfuscus), lead levels in animal samples were below those of their estimated diets, while cadmium was accumulated by several of the investigated species (slugs, earthworms, spiders, ants, and by fungi). Accumulation of Cd by earthworms and slugs already has been reported by IRELAND (1979a, 1979b) and COUGHTREY & MARTIN (1977). In agreement with ROBERTS & JOHNSON (1978), a greater potential for accumulation of cadmium than of lead in terrestrial food webs can be stated. In arthropods, lead is bound in the calcified zones beneath the chitinous exoskeleton (HUNTER & JOHNSON 1982) and therefore relative unavailable to predators, while cadmium is located in soft tissues and taken up by predators. Also in vertebrates, lead is bound in the skeleton (in adult humans 95% of total lead are found in the skeleton, UBA 1976) and therefore immobilized in food chains. Cadmium, however, is accumulated in kidney and liver and available to predators. The distribution of Cd and Pb in vertebrate organisms as documented in this study is supported by several authors (HUNTER & JOHNSON 1982, KREUTZER & KRACKE 1981, ROBERTS & JOHNSON 1978, WARD & BROOKS 1978).

Within the same trophic level, very different cadmium and lead contents were found for the individual species. Of course, the food web scheme shown in Fig.4 and Fig.5 is not complete at all, but it elucidates that Cd and Pb levels in animal samples appear to be rather a function of nourish-physiological attributes of each individual species than of the position in the food web alone.

6. LITERATURE

ANDRESEN,A.M.; JOHNSON,A.H. & SICCAMA,T.G.,1980: Levels of lead,copper, and zinc in the forest floor in the Northeastern United States.
J.Environ.Qual. 9, 293-296
CHMIEL,K.M. & HARRISON,R.M.,1981: Lead contents in small mammals at a roadside site in relation to the pathway of exposure.
Sci.Total Environ. 17, 145-154.
COUGHTREY,P.J. & MARTIN,M.H.,1977: The uptake of lead, zinc, cadmium, and copper by the pulmonate mollusc Helix aspersa Muller, and its relevance to the monitoring of heavy metal contamination of the environment.
Oecologia 27, 65-74.

FANGMEIER,A; STEUBING,L. & GNITTKE,J.,1984: Analyse der Schadstoffbelastung verschiedener Elemente eines Waldökosystems.
Verh.Ges.Ökologie XIV, Hohenheim, in press.
HEINRICHS,H. & MAYER,R.,1980: The role of forest vegetation in the biogeochemical cycle of heavy metals.
J.Environ.Qual. 9, 111-118.
HUNTER,B.A. & JOHNSON,M.S.,1982: Food chain relationship of copper and cadmium in contaminated grassland ecosystems.
Oikos 38, 108-117.
IRELAND,M.P.,1979a: Metal accumulation by the earthworms Lumbricus rubellus, Dendrobaena veneta and Eiseniella tetraedra living in heavy metal polluted sites.
Environ.Pollut. 19, 201-206.
IRELAND,M.P.,1979b: Distribution of essential and toxic metals in the terrestrial gastropod Arion ater.
Environ.Pollut. 20, 271-278.
JOHNSON,A.H.; SICCAMA,T.G. & FRIEDLAND,A.J.,1982: Spatial and temporal patterns of lead accumulation in the forest floor in the Northeastern United States.
J.Environ.Qual. 11, 577-580.
KELLER,T. & ZUBER,R.,1970: Über die Bleiaufnahme und Bleiverteilung in jungen Fichten.
Forstwiss.Centralbl. 89, 20-26.
KREUTZER,W. & KRACKE,W.,1981: Untersuchungen zum Übergang von Blei auf Fleisch und Organe von Rindern unter natürlichen Haltungs- und Fütterungsverhältnissen.
Schriftenr. Bundesminister für Ernährung, Landwirtschaft und Forsten 254: Zum Carry-over von Blei, 28-33.
MARTIN,M.H.; DUNCAN,E.M. & COUGHTREY,P.J.,1982: The distribution of heavy metals in a contaminated woodland ecosystem.
Environ.Pollut. 3, 147-157.
MAYER,R.,1981:Natürliche und anthropogene Komponenten des Schwermetallhaushalts von Waldökosystemen.
Göttinger Bodenkundliche Berichte 70,1-152.
MILLER,W.P. & MCFEE,W.W.,1983: Distribution of cadmium, zinc, copper, and lead in soils of industrial Northwestern Indiana.
J.Environ.Qual. 12, 29-33.
MÜLLER,P.,1982: Experimental bio-monitoring, food-web monitoring and specimen banking.
LEWIS,R.A.; STEIN.N. & LEWIS,C.W.(Edts.):Environmental specimen banking and monitoring as related to banking.
The Hague/Boston/London.
PARKER,G.R.; MCFEE,W.W. & KELLY,J.M.,1978: Metal distribution in forested ecosystems in urban and rural Northwestern Indiana.
J.Environ.Qual. 7, 337-342
PRINZ,B. & SCHOLL,G.,1975: Erhebungen über die Aufnahme und Wirkung gas- und partikelförmiger Luftverunreinigungen im Rahmen eines Wirkungskatasters.
Schriftenr.d.Landesanstalt für Immissions- und Bodennutzungsschutz d. Landes NW 36,62-86.

ROBERTS,R.D. & JOHNSON,M.S.,1978: Dispersal of heavy metals from abandoned mine workings and their transference through terrestrial food chains.
Environ.Pollut. 16, 293-310.
SICCAMA,T.S. & SMITH,W.H.,1980: Changes in lead, zinc, copper, dry weight, and organic matter content of the forest floor of white pine stands in Central Massachusetts over 16 years.
Environ.Sci.&Technol. 14, 54-56.
STEUBING,L.,1985: Pflanzen als Bioindikatoren für Luftverunreinigungen.
Chemie in unserer Zeit 19, 42-47.
STEUBING,L.; GROBECKER,K.H. & KURFÜRST,U.,1980: ZEEMAN-Atomabsorption zur Bestimmung von Schwermetallen in Pflanzen.
Staub-Reinhalt.Luft 40, 537-540.
STEUBING,L.; GNITTKE,J. & GROBECKER,K.H.,1984: Blei- und Cadmiumbelastung eines agrarischen Ökosystems.
Angew.Bot. 57, 1-6.
TURNER,M.A.,1973: Effect of cadmium treatment on cadmium and zinc uptake by selected vegetable species.
J.Environ.Qual. 2, 118-119.
UBA 1976: Luftqualitätskriterien für Blei. UBA-Berichte 3/76.
VAN HOOK,R.I.; HARRIS,W.F. & HENDERSON,G.S.,1977: Cadmium, lead, and zinc distribution and cycling in a mixed deciduous forest.
Ambio 6, 281-286.
VETTER,H.; MÄHLHOP,R. & FRÜCHTENICHT,K.,1974: Immissionsstoffbelastung in der Nachbarschaft einer Blei- und Zinkhütte.
Ber. über Landwirtschaft 52, 327-350.
WARD,N.J. & BROOKS,R.R,1978: Lead levels in sheep organs resulting from pollution from automotive exhausts.
Environ.Pollut. 17, 7-12

STUDIES ON BIOCENOSES, INDIVIDUAL ORGANISMS AND DEPOSITION RATES IN THE EGGE MOUNTAINS, AN AREA HEAVILY AFFECTED BY FOREST DECLINE

Hans-Joachim Ballach
Gesamtverband des deutschen Steinkohlenbergbaus
Postfach 10 36 63
D-4300 Essen 1

Wilhelm Elling
Universität Dortmund
D-4600 Dortmund

Hartmut Greven
Zoologisches Institut der Universität Münster
Hüfferstraße 1
D-4400 Münster

Rüdiger Wittig
Abteilung Geobotanik
Universität Düsseldorf
D-4000 Düsseldorf

ABSTRACT. An overview of a research program performed in the Egge mountains is given. The project comprises monitoring of aerial pollution using plant and animal communities as well as selected organisms in conjunction with chemical measurement (deposition rates, soil analysis). Some findings from a transect at Schwaney are dealt with, in particular those concerning heavy metal pollution (Pb, Zn, Cd). The total and the wet deposition of lead in the open field increase with rising precipitation amount.
    Within deciduous forests along a height gradient in the transect, lead and zinc contents generally increase with the height above sealevel in plant material (moss M n i u m   h o r n u m, foliose lichen H y p o g y m n i a   p h y s o d e s), in animal tissue (earthworm D e n d r o b a e n a   r u b i d a) and in upper soil layers of different soil types.

1. INTRODUCTION

Atmospheric pollution often has significant effects on plant and animal communities. To recognize and monitor such effects, however, an important prerequisite is the knowledge of the "actual state" of the areas under investigation, which often is entirely unknown. Passive and active monitoring of air pollution should be performed by 1) bioindication (possible at the species level but also at the level of biocenoses) as well as by 2) deposition measurements. Both have their specific advantages and disadvantages (Schubert et al. 1985).

Since 1983 we have been studying some biocenoses living near or on ground level (soil fauna, herb layer in beech forests) and deposition rates in North Rhine-Westphalian forests. Additionally, we investigate the effects of "stressors" of anthropogenic origin such as low levels of soil pH or heavy metal depositions on selected organisms in the field and in laboratory experiments with controlled abiotic factors, among others to test their suitability as bioindicators.

A general overview, methods and further data of our research program performed in North Rhine-Westphalian forests were published elsewhere or will be published in the future (Wittig et al. 1985; Ballach et al. 1985; Masuch 1984, 1985; Wittig 1986; Wittig and Werner 1986; Gerdsmeier and Greven in press; Greven in press; Greven et al. in press). Due to the limited space available we report only on some findings from the Egge mountains, one of the most affected areas in North Rhine-Westphalia with respect to spruce decline (Ballach and Brandt 1985).

2. AREAS OF INVESTIGATION

The Egge mountains at the eastern edge of the Westphalian Bight are extending in North-South direction. The height above sea-level ranges between 240 and 468 m.

Our field studies are carried out in the Schwaney forestry within a transect between 315 and 416 m above sea-level. The average annual amount of precipitation increases from about 850 mm at 300 m above sea-level to an average 1100 mm at 400 m.

3. RESULTS AND DISCUSSION

3.1. Atmospheric Pollutants

Considerable pollution concentrations (in particular SO_2) occur in the Egge mountains over a period of hours or even of days (fig. 1). The Egge mountains are dominated by very strong westerly winds, but south-easterly winds also occur frequently. SO_2 concentrations show high values, in particular for south-easterly winds. The concentrations for westerly winds, i. e. from the Ruhr area are lower. Opposite relations were found when evaluating the mass flows (Pfeffer 1985). Referring to NO both the distribution of conentrations as a function of wind direction and the mass flow balances show an increase for westerly winds.

Our samples for deposition measurements (wet and dry depositions and suspended matter) are taken with an automatic wet/dry deposition collector (see Georgii et al. 1980). The sampler has been installed in the open field at 395 m above sea-level in the Schwaney forestry.

Fig. 1: Seasonal variations of monthly average ① and maximum 1/2-hourly concentrations ② of SO_2, O_3, NO and NO_2 in the Egge mountains. The monthly average concentrations of NO are generally less than 10 µg/m³. Measurements from Landesanstalt für Immissionsschutz, Essen (1983-1985).

The cumulative frequency-distribution of pH in precipitation is presented in fig. 2. The 50%-value of the precipitation pH was determined as 4.3. This value is comparable with results from less polluted stations, e.g. Deuselbach (Perseke 1982).

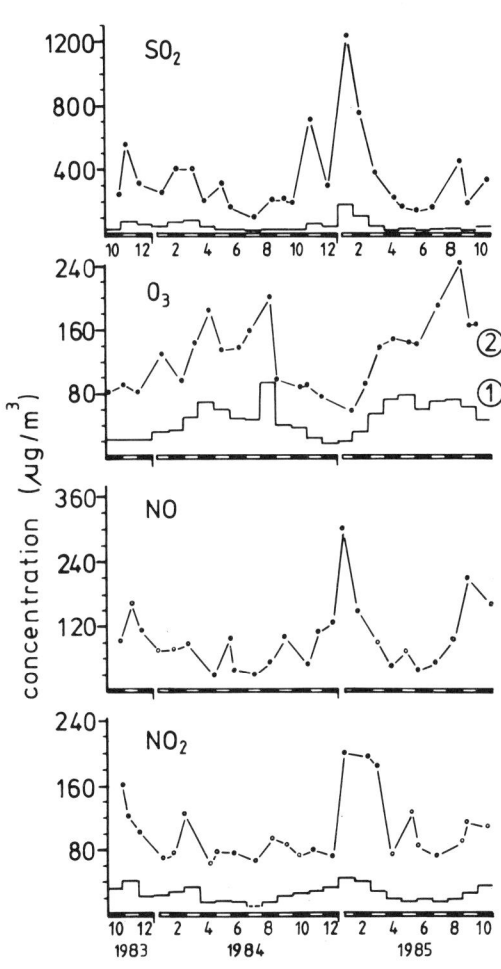

Fig. 2: Cumulative frequency-distribution of pH in open field precipitation (Schwaney forestry). Sampling period: July-December 1985; n=74.

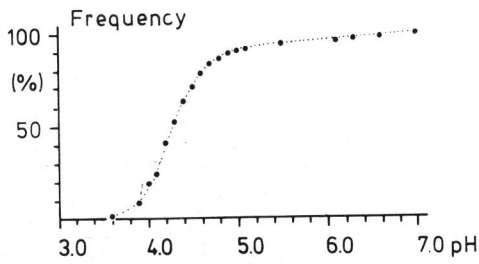

The frequency distribution shows a wide range of pH-values (between 3.6 and 7.0). Results from deposition measurements of sulfate and nitrate, which belong to the main acid producing components in rain, are given in fig. 3.

Fig. 3: Temporal variation of wet (w) and dry (d) depositions of nitrate (N) and sulfate (S).

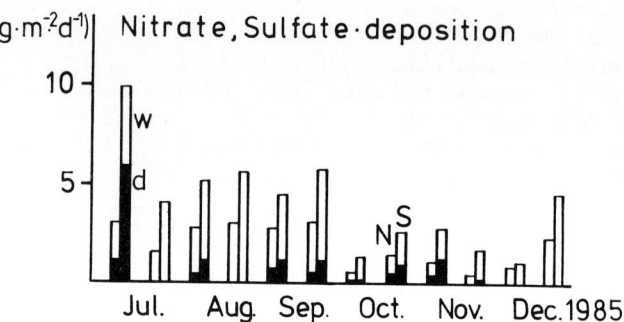

The regional distribution of wet nitrate and sulfate deposition is mainly determined by the precipitation pattern (Perseke 1982). Examinations of snow samples show, among others, increasing nitrate and sulfate depositions in higher regions of the transect (400 m and 416 m) as compared to lower altitudes (315 m and 355 m).

The sulfate contents of snow from January 1986 are shown in fig. 4.

Fig. 4: Sulfate content of snow samples in the transect Schwaney for the period 30 Dec. 1985 -21 Jan. 1986. Date of sampling: 21. Jan 1986.

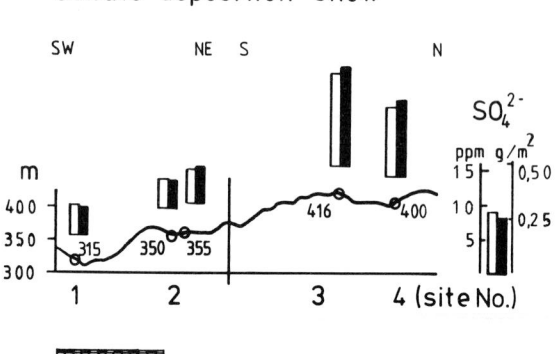

As an example for our deposition measurements of heavy metals, some results obtained for lead are given in fig. 5. Between the amount of precipitation and different types of lead deposition the following coefficients of correlation were found:

r = 0,73 (amount of precipitation/wet deposition)
r = 0,64 ("   "   "   /total deposition)
r =-0,60 ("   "   "   /suspended matter).

The total wet deposition of lead is determined by the precipitation pattern to a large extent (Rohbock 1982). On an average the value of total lead deposition in the open field (Schwaney forestry) appears to be low : 30 µg . m^{-2} . d^{-1} has to be considered as background deposition of lead in the Federal Republic of Germany.

Fig. 5: Precipitation amount, lead deposition and suspended matter in the open field (Schwaney forestry).

ws: water soluble fraction

wi: water insoluble fraction

w: wet deposition

d: dry deposition

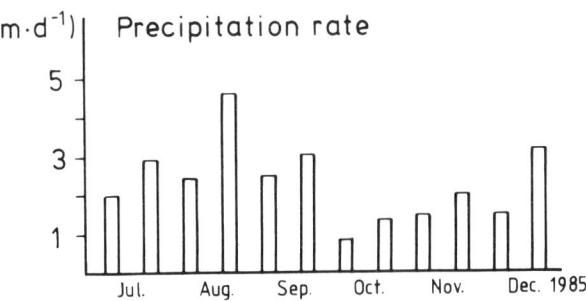
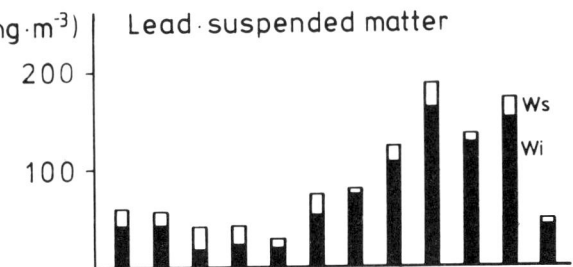
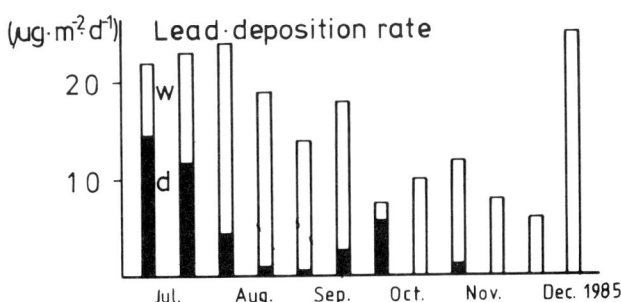

3.2 Animal and Plant Communities

The <u>invertebrate soil fauna</u> is involved in mixing the organic and inorganic constituents of the forest soil and/or in decomposition by consuming dead plant material or the microflora. Therefore the actual state of several zoocenoses on or in the soil of the transect Schwaney has been examined. From Carabidae, Enchytraeidae, Collembola, Lumbricidae and Gastropoda at least a one year study is completed. Some of the zoocenoses (e. g. Collembola, Enchytraeidae) will be studied until 1987.

In the transect Schwaney the number of species of enchytraeids - these oligochaetes are apparently very suitable for bioindication - as well as the dominances of several species show a certain relation between the pH-values of the soil (Ballach et al. 1985). Generally however the period of study is still too short to warrant a more detailed interpretation.

The <u>forest communities</u> of the transect Schwaney belong to the Luzulo-Fagetum (site 1,2 and a small area of site 4) and the Melico-Fagetum (site 2,3,4). In all cases the tree layer consists exclusively of Fagus sylvatica. A shrub layer does not exist. The herb layer of the Luzulo-Fagetum covers 5-10 % (site 1) respectively 10-25 % (site 2). Character species is the acidophilous L u z u l a   l u z u l o i d e s . In the Melico-Fagetum stands the soil is covered by herbs to 50-80 % (site 3), 75-85 % (site 2) and 80-98 % (site 4). Character species are M e l i c a   u n i f l o r a ,   G a l i u m   o d o r a t u m   and H o r d e l y m u s   e u r o p a e u s . The Luzulo-Fagetum under investigation grows on acidic brown earth, the Melico-Fagetum on "Parabraunerde", respectively "Kalkstein-Braunlehm" (table 2). The average soil pH-value ($CaCl_2$) of the A_h-horizon of the Luzulo-Fagetum in the areas not affected by stemflow is 3.4. In the stemflow areas the pH-value ($CaCl_2$) ranges between 2.9 and 3.1. The values for the Melico-Fagetum are 4.3-4.8 in the A_h-Horizon of "normal" forest soil and 3.2-3.4 in the stemflow area.

In order to study possible variations of the plant communities in future, permanent plots have been established at site 2.

Tab. 1: Forest communities and types of soil from the transect Schwaney

Site-No.	Height above sea-level (m)	Forest vegetation	Soil
1	315-320	Luzulo-Fagetum	
2	350-355	Luzulo-Fagetum	acidic brown earth
"	"	Norway Spruce-Forest	
"	"	Melico-Fagetum	"Parabraunerde"
3	410-416	Melico-Fagetum	resp. "Kalkstein-
4	395-400	Melico-Fagetum	Braunlehm"
"	"	(Luzulo-Fagetum)	(acidic brown earth)

Lichens are considered to be among the most pollution sensitive plants, often more sensitive and better indicators of chronic pollution than vascular plants. However, in most cases their use as bioindicators is only an index of general air pollution.

Le Blanc and Rao (1975) concluded that long-term average concentrations of SO_2 between 16 to 79 $\mu g/m^3$ cause chronic injury to epiphytes. At the LIS monitoring station located about 14 km north of the transect an annual average SO_2 concentration of 37.3 $\mu g/m^3$ was measured in 1984. For the two year period from October 1983 to September 1985 an average of 43.7 $\mu g/m^3$ SO_2 was determined (calculated from LIS Monatsberichte 1983-85).

Results of lichen mapping and the indices of atmospheric purity (IAP) for the phorophyte *Quercus robur* in the transect have already been published (Masuch 1984, 1985). The findings suggest that the low and the high level of the transect are more affected than the middle one.

3.3 Heavy Metal Contents of Selected Organisms and Soil

As some plant and animal tissues can reflect the trace metal content of the surrounding environment, such indicators have frequently been used to monitor elemental contamination.

The foliose lichen *Hypogymnia physodes*, which shows a considerable decline along the height gradient in the transect Schwaney (Masuch 1984) and the moss *Mnium hornum* were selected to study heavy metal content.

The arithmetic means and concentration range values of Pb, Zn and Cd in *H. physodes* and *M. hornum* from three different height levels of the transect are given in table 2.

Various plants including mosses and lichens have been tested for their potential to monitor aerial deposition of metals. However, their suitability for such use was found to vary because the heavy metal content of lichens does not necessarily correlate with atmospheric impact. For certain elements an appreciable uptake from the substrate can be expected (de Bruin 1985). Metal uptake of lichens is a passive diffusion process by which different species selectively absorb metals. These are deposited either on the outer surface of the walls of the fungal symbiont or within fungal walls. Heavy metals are immobilized in the fungal partner and do not affect the algal symbiont unless very high concentrations are accumulated (Manning and Feder 1980). Histochemical investigations reveal high metal contents on the surface of *H. physodes* in the transect (Rüther and Greven unpubl.).

Tab. 2: Arithmetic means and concentration range values of metals in
H. physodes and M. hornum from beech
forests (centre of stands) of different heights (315-20 m,
350-55 m and 410-16 m) in the Egge mountains. Five lichens
(Rüther and Greven unpubl.) and three (site 1) respectively
six moss samples from site 2,3 (Wittig and Hillmann unpubl.)
were studied per level. Dates of sampling and analysis:
Nov. 1985 (lichens), Sept. 1985 (mosses).

		Hypogymnia physodes		
Element	Site-No.	Height above sea-level (m)	Mean Concentration (mg/kg dry wt)	Range
Pb	1	315-20	66.64	43.2-91.5
	2	350-55	76.32	59.6-112.8
	3	410-16	71.44	64.2-79.2
Zn	1	315-20	44.88	38.0-57.8
	2	350-55	47.20	37.1-63.9
	3	410-16	50.50	39.6-59.0
Cd	1	315-20	0.208	0.14-0.28
	2	350-55	0.246	0.18-0.31
	3	410-16	0.248	0.20-0.32

		Mnium hornum		
Element	Site-No.	Height above sea-level (m)	Mean Concentration (mg/kg dry wt)	Range
Pb	1	315-20	74.6	70.2-80.1
	2	350-55	126.4	96.3-202.9
	3	410-16	145.2	116.1-234.2
Zn	1	315-20	87.4	80.2-101.7
	2	350-55	142.9	101.9-163.3
	3	410-16	147.7	136.8-155.1
Cd	1	315-20	0.87	0.71-1.02
	2	350-55	0.96	0.77-1.17
	3	410-16	1.24	1.05-1.49

Mosses are considered to be the most effective heavy metal accumulators among green plants (Thomas 1983). Generally, the heavy metal contents (Pb, Zn, Cd) found in M. hornum of the Schwaney forestry are within the range occurring in industrial regions of Great Britain (Goodman and Roberts 1971; Ellison et al. 1976) and polluted areas of Scandinavia (Rühling and Tyler 1973). It should be emphasized that M. hornum belongs to the endohydric mosses (Buch 1947) so that an uptake of metals from soil cannot be entirely excluded.

As earthworms accumulate several heavy metals from contaminated soils particularly with their chloragogenous tissue, they may be of use as indicators for this kind of pollution. Martin and Coughtrey (1982) conclude that earthworms may be useful monitors of lead and cadmium in soil, while in the case of zinc further research is required before such a conclusion can be drawn. The uptake of metals is influenced by soil pH. The arithmetic means and concentration range values of Pb, Zn and Cd in the chloragogenous tissue of Dendrobaena rubida, a common earthworm in the transect, for different height levels, are listed in table 3. The heavy metal content of soils are shown in table 4 for comparison.

Tab. 3: Arithmetic means and concentration range values of metals in the chloragogenous tissue of Dendrobaena rubida. Five earthworms were taken per level. Date of sampling: November 1985.

Element	Site-No.	Height above sea-level (m)	Mean Concentration (mg/kg dry wt)	Range
Pb			22.156	18.42-27.80
Zn	1	315-320	18.032	13.65-21.95
Cd			0.094	0.06- 0.12
Pb			24.082	20.70-24.74
Zn	2	350-355	25.478	21.27-28.92
Cd			0.090	0.07- 0.11
Pb			26.316	21.60-31.97
Zn	3	410-416	28.464	27.14-31.24
Cd			0.100	0.08- 0.12

Tab. 4: Cd-, Pb- and Zn content of soil (mg/kg dry wt) in beech forest of the transect Schwaney (Sept. 1985). Total contents: First column, available contents: Second column.

		Site 1		Site 2		Site 3		Site 4	
Pb	0-4 cm	91.69	8.24	117.20	7.26	119.25	7.12	117.26	8.92
	4-8 cm	90.90	7.88	104.70	6.68	109.35	7.28	111.34	8.23
Zn	0-4 cm	197.60	11.43	178.65	9.56	184.30	9.08	202.70	11.26
	4-8 cm	195.57	8.95	164.32	8.42	181.76	8.98	205.78	11.16
Cd	0-4 cm	0.778	0.195	0.898	0.221	0.986	0.269	1.005	0.198
	4-8 cm	0.798	0.176	0.841	0.157	0.898	0.234	0.888	0.276

Studies on earthworms (L u m b r i c u s t e r r e s t r i s) kept in artificial soil (OECD) contaminated with different lead concentrations not only suggest a strong relationship between the lead concentration in animal tissue and soil (see also field data obtained by Morgan 1985), but also show that the nucleus volume of chloragogenous tissue is significantly reduced depending on the rate of contamination (Rüther and Greven in prep.).

ACKNOWLEDGEMENTS

This project was supported by the Gesamtverband des deutschen Steinkohlenbergbaus, Essen, West Germany. We are indebted to W. Rüther, J. Schoch and R. Hillmann for generously adding some of their results to this report.

REFERENCES

Ballach, H.-J., Greven, H. and R. Wittig (1985): Biomonitoring in Waldgebieten Nordrhein-Westfalens. Überblick und erste Ergebnisse. Staub - Reinhalt. Luft 45, 567-573.

Ballach, H.-J. and J. Brandt (1985): Verteilung der Fichtenschäden in Nordrhein-Westfalen. Staub - Reinhalt. Luft 45, 1-6.

Buch, H. (1947): Über die Wasser- und Mineralstoffversorgung der Moose. Soc. Scient. Fennica, Comment. Biol. 9.

DeBruin, M. (1985): Epiphytic lichens as indicators for heavy metal airpollution: What do they reflect? T. D. Lekkas (ed.), Heavy Metals in the Environment, 1, 359-361. Int. Conf. Athens, Sept. 1985.

Ellison, G., Newham, J., Pinchin, M. J. and I. Thompson (1976): Heavy metal content of mosses in the region of Consett (North East England). Environ. Poll. 11, 167-174.

Georgii, H.-W., Gravenhorst, G., Perseke, C. and E. Rohbock (1980): Untersuchung über die trockene und feuchte Deposition von Luftverunreinigungen in der Bundesrepublik Deutschland. Bericht - Aufträge des Umweltbundesamtes, Oktober 1980.

Gerdsmeier, J. and H. Greven: Zur Kenntnis der Collembolenfauna des Eggegebirges. Abh. Landesmus. Naturk. Münster, in press.

Goodmann, G. T. and T. M. Roberts (1971): Plants and soils as indicators of metal in the air. Nature 231, 287-292.

Greven, H.: Vermehrung epidermaler Schleimzellen als Antwort von Lumbriciden und Gastropoden auf Streßsituationen. Verh. Ges. Ökol., in press.

Greven, H., Bettin, Ch., Reichelt, R. and U. Rüther: Die Wirkung von Säurestreß auf Lumbricus terrestris (Lumbricidae, Oligochaeta). Methodik und erste Ergebnisse. Verh. Ges. Ökol., in press.

Landesanstalt für Immissionsschutz des Landes Nordrhein-Westfalen, Essen, (1983-1985): Monatsbericht über die Luftqualität an Rhein und Ruhr.

Le Blanc, F. and D. N. Rao (1975): Effects of air pollutants on lichens and bryophytes. J. B. Mudd and T. T. Kozlowski (eds.), Responses of Plants to Air Pollution, 237-272. Academic Press, New York.

Manning, W. J. and W. A. Feder (1980): Biomonitoring air pollutants with plants. K. Mellanby (ed.), Applied Science Publishers.

Martin, M. H. and P. J. Coughtrey (1982): Biological monitoring of heavy metal pollution. K. Mellanby (ed.), 142 pp. Applied Science Publishers, London.

Masuch, G. (1984): Besiedlungssukzessionen der Flechte Hypogymnia physodes (L.) Nyl. entlang eines Höhengradienten im Eggegebirge. Staub - Reinhalt. Luft 44, 492-496.

Masuch, G. (1985): Flechtenkartierung entlang eines Niederschlagsgradienten im Eggegebirge. Staub - Reinhalt. Luft 45, 573-578.

Morgan, J. E. (1985): The interaction of exogenous and endogenous factors on the uptake of heavy metals by the earthworm Lumbricus rubellus. T. D. Lekkas (ed.), Heavy Metals in the Environment, 1, 736-738. Int. Conf. Athens, Sept. 1985.

Perseke, C. (1982): Composition of acid rain in the Federal Republic of Germany - Spatial and temporal variations during the period 1979 - 1981. H. W. Georgii and J. Pankrath (eds.), Deposition of Atmospheric Pollutants, 77-86. D. Reidel Publishing Company.

Pfeffer, H. U. (1985): Immissionserhebungen in quellfernen Gebieten Nordrhein-Westfalens. Staub - Reinhalt. Luft 45, 287-293.

Rohbock, L. (1982): Atmospheric removal of airborne metals by wet and dry deposition, 159-171. H. W. Georgii and J. Pankrath (eds.), Deposition of Atmospheric Pollutants, 159-171. D. Reidel Publishing Company.

Rühling, A. and G. Tyler (1973): Heavy metal deposition in Scandinavia. Water, Air Soil Pollution 2, 445-455.

Schubert, R. (ed.) (1985): Bioindikation in terrestrischen Ökosystemen. Gustav Fischer Verlag, Stuttgart, 327 pp.

Thomas, W. (1983): Über die Verwendung von Pflanzen zur Analyse räumlicher Spurensubstanz - Immissionsmuster. Staub - Reinhalt. Luft 43, 141-148.

Wittig, R., Ballach H.-J. and C. J. Brandt (1985): Increase of number of acid indicators in the herb layer of millet grass-beech forest of the Westphalian Bight. Angew. Bot. 59, 219-232.

Wittig, R. (1986): Veränderungen in der Krautschicht von Buchenwäldern. IMA - Querschnittsseminar, Belastungen und Schäden auf Ökosystemebene Umweltbundesamt-Texte, in press.

Wittig, R. and W. Werner (1986): Untersuchungen zur Belastungssituation des Flattergras-Buchenwaldes in der Westfälischen Bucht. Düsseldorfer Geobot. Kolloq. 3, in press.

DEPOSITION/CANOPY-INTERACTIONS IN TWO FOREST ECOSYSTEMS OF
NORTHWEST GERMANY

E. M a t z n e r;
Forschungszentrum Waldökosysteme/Waldsterben,
Universität Göttingen, Büsgenweg 2,
3400 Göttingen

Abstract

From the flux balance of the forest canopy the rates of total deposition (precipitation- + interception-deposition) in two forest stands are calculated for the elements H^+, Na, K, Ca, Mg, Mn, Al, Fe, S, Cl and N. Special emphasis is given to the seasonal patterns of deposition.

The deposition of protons results in a stress to the buffer systems of the leaves. The buffering of protons on the surfaces of leaves and bark is connected with the leaching of cations, mainly Ca. The rates of proton buffering and the consequences are discussed as well as the seasonal patterns and rates of cation leaching.

Determination of the rates of total deposition of N from the flux balance of the canopy proved to be difficult because significant amounts of N may be assimilated within the canopy.

Stands and Methods

The data used in this paper stem from 2 different stands in North Germany located on the plateau of the Solling mountains (500 m NN) and are represented by a 135 years old beech stand (FA Neuhaus, Abt. 51) and by a neighbouring stand of Norway spruce, 100 years old (FA Dassel, Abt. 28). Both stands are growing on podzolic brown earth soils derived from loess, overlaying sandstone bedrock. Further details about these stands are given by ELLENBERG (1972) and SEIBT (1981). Soil chemical data and the effects of deposition of air pollutants on the soils are discussed by ULRICH et al. (1979), MATZNER and ULRICH (1981) and by ULRICH and MATZNER (1983).

The rates of deposition and the turnover of ions within the forest canopy were calculated by using the method developed by ULRICH (1983). For calculation ULRICH uses the fluxes of elements with bulk precipitation (so called precipitation-deposition = PD) and throughfall (including stemflow which is of importance for beech only).

The sampling is done by 15 standard rain collectors made of plexiglass on each site. The solutions were taken weekly and mixed for monthly samples. During wintertime 9 plastic buckets were used to collect snow samples. The samples were analysed for H^+ (glass-electrode), Na, K, Ca, Mg, Mn (AAS), Fe, Al, PO_4, NH_4, NO_3 (colorimetric methods), SO_4 and Cl (titrimetric methods).

Data on canopy-deposition in these stands for heavy metals are given by MAYER (1983) and for organic pollutants by MATZNER (1984).

Results

I Rates of Deposition

The rates of precipitation deposition (PD = bulk precipitation) and of total deposition of the alkali-earthalkali-elements and of H^+, Fe, Mn, S and Cl are presented in Tab. 1.

Looking at the data of the Solling stands, the strong influence of the forest canopy on the rates of total deposition becomes evident. The rates of interception deposition (ID = Adsorption of gases, aerosols and droplets) are significantly higher for the spruce stand than for beech (ID calculated as difference of PD and total deposition). This is caused by the larger filtering area of the spruce canopy especially during winter.

The relation of total deposition/PD in both stands indicates that the rates of interception deposition of H^+ and S are high as compared to the other elements. This effect can be attributed to the deposition of SO_2 with subsequent formation of sulfuric acid in the water films on the leaves, bark and needles. Furtheron the adsorption of unbuffered acid droplets may give the same result.

The range of the annual rates of deposition given in table 1 which was observed during the 15 years of investigation emphazises the importance of long lasting measurements for determining the deposition to a forest stand. The variation in total deposition from year to year may exceed 100% whereas the variation in precipitation deposition is less. However, an overall trend was not evident. The lowest rates of deposition of H^+ and S throughout the measuring period was recorded in the warm/dry years of 1982 and 1983. The conditions for

Tab. 1: Rates of deposition in the Solling area (kg · ha^{-1} · a^{-1}; $\bar{x}$ 1969-83) '(range)

		H	Na	K	Ca	Mg	Fe	Mn	Al	SO$_4$-S	Cl
Precipitation deposition	$\bar{x}$	0.79 (0.61-1.10)	7.9 (4.9-12.1)	3.7 (2.4-5.6)	10.0 (6.5-21.8)	1.8 (1.3-3.9)	0.8 (0.2-1.1)	0.4 (0.1-0.9)	1.1 (0.6-2.1)	23.4 (19.6-27.2)	16.8 (10.5-25.5)
Total deposition beech	$\bar{x}$	1.90 (1.13-2.62)	14.2 (9.5-18.2)	6.7 (4.9-9.9)	17.5 (11.4-32.0)	3.0 (2.1-3.8)	1.5 (0.3-2.4)	0.7 (0.2-1.4)	2.0 (0.9-3.2)	50.3 (39.1-66.0)	32.7 (24.2-40.9)
Total deposition spruce	$\bar{x}$	3.81 (2.68-5.26)	17.4 (9.0-25.4)	8.1 (5.7-11.9)	22.1 (13.2-37.9)	4.2 (2.4-6.3)	1.8 (0.4-3.1)	0.9 (0.2-1.8)	2.4 (1.0-4.5)	85.3 (68.8-108)	39.4 (27.5-54.3)

the deposition of SO_2 are obviously unfavourable in warm/dry years as compared to wet/cool years. This may be caused by low relative humidity and the lack of water films during drought. The deposition of H^+ may be further reduced due to the high dust concentrations in the air causing subsequent buffering during 1982 and 1983. The low rates of deposition of H^+ and S in forests which were found in various parts of Germany in 1982 and 1983 therefore presumably underestimate the long term input.

Figure 1

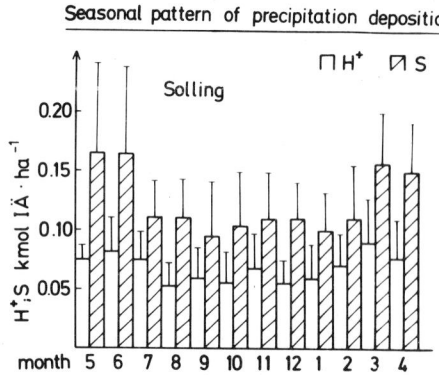
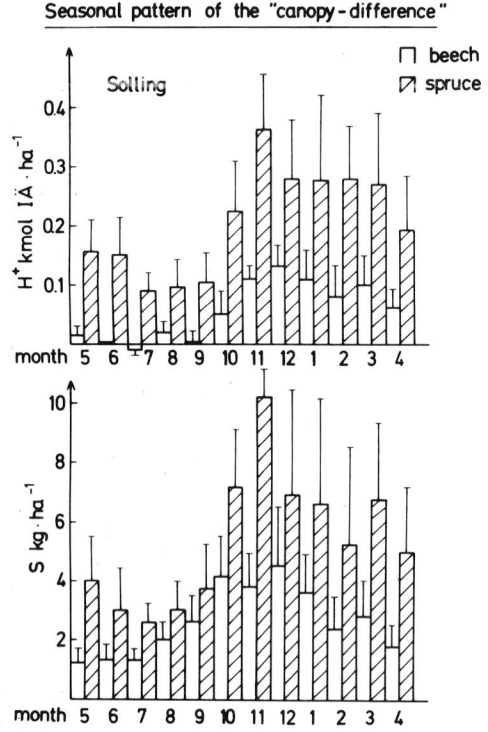

Further information about the mechanisms of deposition are available from the seasonal pattern of the rates of deposition. Figure 1 therefore shows the mean monthly values ($\bar{x}$ 13 years from 1969-1981 ± 95% conf.interval) of the rates of precipitation deposition of H^+ and S and the mean monthly values of the difference between throughfall and PD. This difference is called "canopy difference" = CD and corresponds to the rate of interception deposition in the case of S. According to ULRICH (1983) interception deposition is defined as adsorption of gases, aerosols and droplets.

In the case of H^+ the CD does not correspond to the rate of interception deposition since protons are partly buffered on the plant surfaces. This process will be discussed in detail later. The buffering of protons within the canopy will presumably influence the seasonal pattern of the CD only slightly for spruce, while the low values observed for beech during the vegetation period are partly caused by proton buffering and not by low rates of interception deposition of H^+. Calculations of the rates of H^+-buffering were only done on a yearly base. That is why no correction of the CD of H^+ is possible for monthly values so far.

The rates of PD of H^+ are about the same throughout the season, while a slight maximum of S deposition is found during the summer.

The CD of S and of H^+ are about the same throughout the season, while a slight maximum of S deposition is found during the summer.

The CD of S and of H^+ shows a pronounced seasonal pattern indicating high rates of interception deposition during the winter. One mechanism causing an increase in the rates of interception deposition during winter may be the adsorption of acid droplets as indicated by the occurrence of fog. The mean number of days with fog is given in fig. 1. The maximum is also found during winter. Especially fog arising from long range transport of clouds in the higher elevated sites may be enriched with pollutants, mainly H^+ and S (SCHRIMPFF 1983, WISNIEWSKI 1982). That is why the interception of fog may play a significant role in determining the rates of deposition in higher elevations. This was indicated already by ULRICH et al. (1979).

II. Turnover of elements within the canopy

As mentioned above, a part of the total deposition of H^+ is buffered during canopy passage. If the rate of interception deposition is low one can find a net-reduction of the H^+ load in throughfall as compared to the precipitation deposition (ULRICH et al. 1973, CRONAN and REINERS 1983).

To keep the electroneutrality in the plant and the solution, the plant has to take up an amount of anions from precipitation that is equivalent to the protons buffered, or the buffering must be connected with the leaching of cations from the canopy. With regard to the amounts of cations available for leaching, Ca, Mg, K and Mn may be of importance in this respect.

Table 2 gives an overview of the leaching and buffering processes within the canopy of the Solling stands by presenting the mean values and the range observed during the measuring period. The rates of proton buffering are calculated as difference between total deposition and throughfall (ULRICH 1983). The same holds for the leaching rates of K, Ca, Mg and Mn.

Table 2: Turnover of elements in the forest canopy: H^+ buffering and leaching (kmol IÄ $\cdot$ ha^{-1} $\cdot$ a^{-1}; x 1969-83 (range))

SOLLING

	Beech	Spruce
1 H^+-buffering of the canopies	0.57 (0.01-1.13)	0.71 (0.0 -1.59)
leaching of Ca^{2+}	0.34 (0.0 -0.80)	0.56 (0.0 -0.84)
leaching of K^+	0.52 (0.21-0.80)	0.52 (0.35-0.88)
leaching of Mn^{2+}	0.11 (0.09-0.14)	0.16 (0.12-0.24)
leaching of Mg^{2+}	0.07 (0.0 -0.17)	0.05 (0.0 -0.13)
2 Σ leaching cations	1.04	1.29
3 leaching of dissolved organic anions (x 1971-83) 1)	0.34 (0.03-0.67)	0.40 (0.0 -1.31)
4 (1 - 2 + 3)	-0.13	-0.18

1) calculated from the cation/anion balance of bulk-precipitation and throughfall

The leaching of cations from the canopy may also be accompanied by the leaching of dissolved organic anions. The flux of organic anions is calculated from the cation/anion balance of the elements measured in PD and throughfall. The deficit of anions (= amount of organic anions) is, according to table 2, higher in throughfall than in the PD. That is why a positive value is given for the leaching of organic anions from the canopy of these stands.

The mean annual rates of H^+-buffering amount to about 25% of the total H^+-deposition for the beech stand (0.58 keq $\cdot$ ha^{-1}) and to about 20% of the total deposition (0.73 keq $\cdot$ ha^{-1}) for the spruce stand. The whole year needle cover and the higher load of acidity may explain the difference between beech and spruce.

The cation leached most (when expressed in keq) is K for beech, followed by Ca, Mg and Mn. K and Ca leaching is about the same in the spruce stand. The range of values given for both sites indicates a strong annual variability which was also discussed by ULRICH (1983) in detail.

Total amount of cations leached have been calculated as 1.02 (beech) and 1.32 (spruce) keq $\cdot$ ha^{-1} $\cdot$ a^{-1}. The leaching can not be totally balanced by the buffering of protons since leaching of cations is twice the buffering rate. Taking the leaching of organic anions into account, the processes: Leaching of cations, buffering of H^+ and leaching of anions are nearly balanced. The remaining difference in the charge balance can be attributed to different processes. The calculation of the rates of interception deposition was done assuming no leaching of mineral anions. The leaching of about 2-3 kg S $\cdot$ ha^{-1} $\cdot$ a^{-1} (which is very little compared to the fluxes with throughfall) would already be sufficient for balancing. ULRICH et al. 1979 calculated leaching of S from senescent leaves to be about 3.3 kg S $\cdot$ ha^{-1} $\cdot$ a^{-1}.

Furtheron the charge balance may be adjusted by the uptake of other cations beside H^+ into the leaf. The cation taken up may be NH_4^+. Again the uptake of 2-3 kg NH_4^+-N per ha and year would be sufficient to balance the charge deficit.

Before discussing the possible uptake of N by the leaves the process of H^+-buffering and the connected cation leaching should be considered in detail. Ca is taken as an example because Ca may be the dominating cation leached since it is mostly found in the cellwalls of plant tissues.

According to ULRICH (1983) the buffering reaction can be described by three steps:

During the first reaction Ca^{2+} is exchanged from the cellwall by H^+ and reaches the soil together with SO_4^{2-}. The protonated exchange sites of the cellwall are then recharged by $Ca^{2+} + 2\ HCO_3^-$. The third reaction clearly shows the consequences of proton buffering in the canopy: During the uptake of Ca from the soil solution an equivalent amount of protons is released. That is why the buffering in the canopy does not reduce the acid load resulting from deposition. But the amount of protons buffered in the canopy causes a stress to the buffer systems of the root surrounding parts of the soil.

The rates of this process may be influenced by the atmospheric load of protons and by the soil conditions. In well buffered soils without reduction of Ca-uptake by Al-ions in the soil solution, a large part of the acid load may be buffered on the leaves (ULRICH and MATZNER 1983) resulting in a strong acidification of the rhizosphere in such stands. On sites that are already strongly acidified the reduced uptake of Ca may become limiting for the buffering process in the canopy.

Increased cation leaching caused by ion exchange in the leaves and needles may contribute to nutrient deficiency as is often reported in connection with the decline of spruce and fir (ZECH and POPP 1983, ZÖTTL and MIES 1983). As mentioned above, the reduction of the cation uptake due to high amounts of Al-ions which was found by ROST-SIEBERT (1983) and HÜTTERMANN (1983) will enhance this effect.

The rates of leaching of K and Mg calculated for the beech stand correspond to the difference between the amount of elements stored in the green leaves and those found in the litterfall (MATZNER et al. 1982). The change in the element storage of the leaves throughout the season therefore can be explained by leaching. As the element storage of litterfall is higher than in the green leaves for Ca, while it is about equal for Mn, the leaching of Ca and Mn can not be attributed to the changing element contents of the leaves throughout the growing season.

The annual leaching of K in the spruce stand amounts to about half of the total storage within the needle biomass. The ratios for Ca and Mg are 1/3 and for Mn 1/4.

Regarding the seasonal pattern of the canopy-difference (CD) for other elements as was already done for H^+ and S, gives further information about the time and rates of leaching. Figure 2 therefore shows the seasonal development of the CD for K, Ca, Mg, Na, NH_4 and NO_3 found in the Solling stands. Rates are given as mean values 1969-1981 ± 95% conf.-interval.

Figure 2:

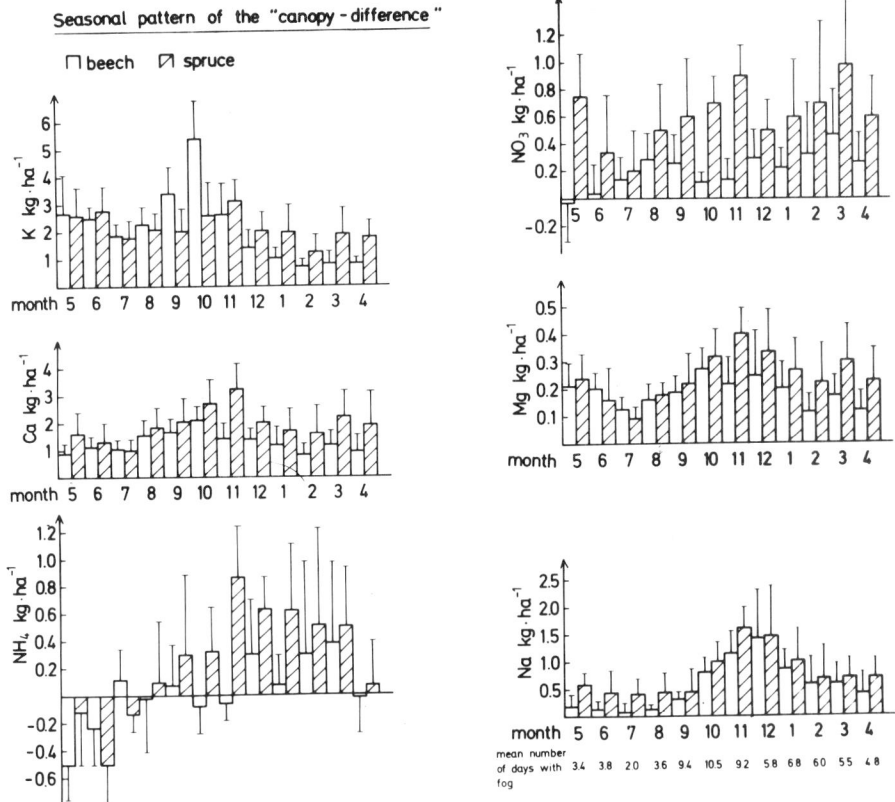

Because the leaching of Na from leaves and needles can be neglected, the seasonal pattern of Na in the CD is therefore identical to the seasonal pattern of interception deposition. Thus the CD of Na can be taken as a background information when evaluating the CD of these elements which are not subjected to gaseous deposition.

The pattern of the CD of K differs significantly when compared to Na especially in the beech stand. Low rates of CD of K in winter period (month 1-4), presumably resulting only from deposition, are followed by a strong increase in its amounts in May at the time of leaf outbreak and the beginning of leaching. The CD of K then keeps constant during month 5-8 and increases again in autumn throughout the period of senescence before litterfall. The leaching of K from senescent leaves is also reported by CRONAN and REINERS (1983).

The pattern of the CD of K in the spruce stand is much more leveled. However, there is an increase in May and June when compared to month 1 to 4. High rates found in October and November may be caused by interception deposition indicated by the behaviour of Na, S and Ca.

Despite the fact that Ca is leached in considerable amounts from the tissues as a consequence of proton buffering, the seasonal pattern of the CD is similar to Na. The reason for this is the high rate of interception deposition of Ca. The ratio of interception deposition/leaching of Ca is about 3:1 for beech and 2:1 for spruce. Therefore the seasonal pattern of the CD for Ca necessarily equals the one of Na. The ratio of interception deposition/leaching for K is about 0.3 and subsequently the effects of leaching on the pattern of CD are easily evident.

The above mentioned ratio for Mg is about the same as for Ca in the beech stand. Nevertheless a pronounced increase of the CD is found for beech in May which must be attributed to leaching. Mg leaching decreases during the summer but, according to K, increases again in autumn. The CD of spruce gives no information about the seasonal pattern of Mg-leaching since the ratio of deposition/leaching is 7:1.

III N-input

To determine the rates of interception deposition and total deposition of N proved to be difficult because several processes may occur in the canopy that cannot be quantified, e.g. the assimilation of N by the plant or by canopy epiphyts. Furtheron one cannot account for hidden turnover of the various N-forms NH_4^-, NO_3^-, N_{org}, NO_x, N_2 und NH_3 in the canopy and in the sampler. Nitrification of NH_4^+ in the canopy or the sampler would have a strong effect on the pH since per mol NH_4^+ 2 mols of H^+ are produced. The importance of this process for the throughfall pH was emphazised by CHEN et al. (1983) in their model on throughfall chemistry.

To sum up, there are a lot of problems associated with the calculation of N deposition. However, an estimate of the N deposition to the stands observed should be made based on the flux measurements. The temporal pattern of the N fluxes with throughfall is characterized by an increase from 1969 to 1976. From 1976 onwards the N fluxes are more or less constant (MATZNER et al. 1982).

The annual fluxes of NH_4, NO_3 and N_{org} with throughfall exceed those with bulk precipitation in case of beech and spruce. The increase is most obvious for N_{org} (MATZNER et al. 1982).

Parallel to other elements investigated the N-fluxes with throughfall under spruce are again higher than under beech.

N may be deposited in the form of particles or, similar to S, by adsorption of gases (NH_3, NO_x, HNO_3-vapour). Assuming that all N deposited will be washed of by precipitation and is measured in the throughfall, the ratio of total deposition/precipitation deposition of NH_4, NO_3 would be 1.3/1.2 for beech and 1.4/2.2 for spruce. These ratios are low when compared to other elements, e.g. for Na which is only deposited in the form of particles the ratio is 2.0 (beech) and 2.4 (spruce). For S the ratios of 2.2 (beech) and 3.6 (spruce) were found. From these data one can conclude that parts of the N deposition do not reach the soil with throughfall but are presumably assimilated in the canopy. This conclusion is also confirmed by the seasonal pattern of the CD of NH_4 and NO_3 given in figure 2. Although the annual NH_4-flux with throughfall of both stands is higher than with bulk precipitation, the CD becomes negative during the summer, indicating NH_4 assimilation in the canopy. No negative values of the CD are found for NO_3 but the CD is rather low during summer and the process of nitrification of NH_4 with subsequent leaching of NO_3 is neglectable. High rates of interception may yield in positive values despite NO_3-assimilation. High rates of deposition of HNO_3-vapour were found in the Solling stands by LINDBERG (pers.comm).

The rates of interception deposition (ID) of N should be estimated roughly by applying the ratio of ID/PD for Na (see ULRICH 1983). This estimate assumes that Na-containing aerosols are deposited in the same amount as N-containing and it may however underestimate the deposition of N because gaseous is neglected.

Estimating the total deposition as described above allows the calculation of N assimilation in the canopy by subtracting the N-flux with throughfall. This results in a rate of N assimilation in the canopy of about 8 kg $\cdot$ ha^{-1} $\cdot$ a^{-1} in both Solling stands, according to 20-25% of the N-flux with throughfall. Since there are a lot of uncertainties about the turnover of N-forms in the canopy no data can be given of the form of N-uptake.

The fluxes of N with throughfall therefore underestimate the actual N-input. Accepting the magnitude for N-uptake in the canopy calculated above, the total annual N-input is about 40 (beech) to 50 (spruce) kg N $\cdot$ ha^{-1} and is equal to the total N demand of the spruce stand and provides 70% of the N demand of the beech stand. Both stands no longer depend on N-mineralization for their N-uptake. However, there are no N-losses with seepage water in the beech stand while the NO_3 is leached from the soil of the spruce stand completely. Both ecosystems are accumulating significant amounts of N in the organic top layer (ULRICH et al. 1980). N may also accumulate in the mineral soil, but, to quantify the rates of total deposition from the inventory studies of the mineral soil and the output data is not promising because large storage are compared with small rates of change.

Discussion

The deposition of acidity and its effects on the element cycling in forest canopies have been the subject of several papers. The process of H^+ buffering in the canopy by ion exchange was mentioned by e.g. COLE and JOHNSON (1977), CRONAN and REINERS (1983) and by ULRICH (1973, 1983, 1983a) but was quantified only in the papers of ULRICH. CRONAN and REINERS distinguish between two reactions involved, ion exchange and the protonation of BRØNSTEDT bases. This concept is similar to the one put forward by ULRICH (1983a) which has been described in this paper. CRONAN and REINERS (1983) and ULRICH (1983) emphazise that the H^+ buffering in the canopy does not reduce the acid load of the soil. From the ecological point of view the acidity buffered in the canopy effects a rather critical space of the system: the apoplast of the root cortex and the rhizosphere. Since the protons produced during ion uptake (cation uptake > anion uptake, MATZNER and ULRICH 1983) also adds to the proton load on the root cortex and the root surrounding soil, one can assume strong acidification of this space. The possible impact of this process via soil acidification and acid toxicity on the stands may be evaluated at first by pH measurements close to roots. The acidification of the rhizosphere of agricultural crops has been demonstrated by MARSCHNER and RÖMFELD (1983) and by SCHALLER and FISCHER (1985). However, the importance of various processes for acidification may differ when comparing agricultural crops and forests. Unfortunately no measurements on the pH close to roots are available in forest stands under field conditions so far.

The effects of proton buffering in the canopy accompanied by the increased leaching of the cations, was the subject to experiments carried out by HORNTVEDT et al. (1980). They report increasing leaching of Ca, Mg and K by raising the load, but did not find changes in the needle content of these elements indicating a quick replacement by ion uptake.

The calculated rates of leaching of K and Mg in the beech stand correspond to the difference of the element content in green leaves and in litterfall. Changing element contents from green to senescent leaves were studied by STAAF (1982) for N, P, K, S, Ca and Mg. STAAF attributed the decrease in element content of all elements to translocation into perennial parts. In the case of K and Mg this cannot be confirmed for the Solling beech stand.

Discussing the N-input by deposition it was mentioned that the N-flux with throughfall underestimates the actual input since clear indications for N-uptake in the canopy were found. The uptake of nutrients by leaves has been known for long time and is practically applied by using leaf-fertilization (BOYNTON 1954). However, no data about the rates of uptake following leaf fertilization of forest stands are available. The rates of N-uptake calculated for the stands under investigation of about 8 kg N/ha therefore can not be compared with the literature data. The process of N-uptake in the canopy was also studied by MILLER and MILLER (1980), but they only discuss changes of N concentrations in precipitation and throughfall over a short period of time without quantifying the uptake rate. The paper of LANG et al. (1976) is focussing on the influence of epiphytic lichens on the throughfall chemistry in a stand of Douglas fir. They report temperature depending uptake of N in the canopy (maximum during summer) and the turnover of mineral to organic N. The organic N is subsequently determined in the throughfall.

The results found in the Solling are in good agreement with the experiments done by LANG et al. when the seasonal pattern of the CD of NH_4^+ and the proportion of organic N in throughfall are considered. However, the biomass of lichens in the Solling stand certainly is much lower than the stand described by LANG, but N turnover may be also possible by bacteria and algae on the leaves.

The long term effects of the present N input in forest ecosystems are only poorly understood. An increase of the increment has been shown for many stands (SEIBT 1981). This increase was followed during the last decades by a drastic reduction of the increment caused by the dominating effects of acid deposition and other pollutants. MEYER (1984) attributed the root decline observed in damaged forest stands of Germany to a disturbance of the mycorrhiza by high levels of N deposition. Evaluation of this hypothesis must be the aim of future research as well as the effects of N uptake by leaves on the ionic status of the leaf and the rates of ion uptake by roots.

LITERATURE

Boynton, D. 1954: Nutrition by foliar application. Ann.Rev.Plant Physiol. 5, 31-54

Chen, C.W., M. Asce, R.J.M. Hudson, S.A. Gherini, H.D. Dean and R. Goldstein 1983: Acid rain model: Canopy module. J.Evir. Engeneering 109, 585-603

Cole, D.W. and Johson 1977: Atmospheric additions and cation leaching in a Douglas Fir ecosystem. Water Resources Research 13, 313-317

Cronan, C.S. and W.A. Reiners 1983: Canopy processing of acidic precipitation by coniferous and hardwood forests in New England. Oecologia (Berlin) 59, 216-223

Ellenberg, H. (ed.) 1972: Integrated experimental ecology, methods and results of ecosystem research in the German-Solling-Project. Ecological Studies 2, Springer Verlag

Glatzel, G. 1983: Die Messung der Deposition langzeitwirksamer Luftschadstoffe in Wäldern. Inst.f.Forstökologie, Universität Wien

Horntvedt, R., Dollard, G.H., Joranger, E. 1980: Effects of acid precipitation on soil and forest II. Atmospheric-vegetation interactions. In: Ecological Impact of Acid Precipitation, Drabløs, D., Tollan, A. (eds.) SNSF-project Oslo-Ås, p. 192

Hüttermann, A. 1983: Auswirkungen saurer Deposition auf die Physiologie des Wurzelraums von Waldökosystemen. Allg.Forstz., 663-664

Lang, G.E., Reiners, W.A. and Heier, R.K. 1976: Potential alteration of precipitation chemistry by epiphytic lichens. Oecologia (Berlin) 25, 229-241

Marschner, H. and V. Römfeld 1983: In vivo measurements of root-induced pH changes at the soil root interface: Effect of plant species and nitrogen source. Z.Pflanzenphysiol. Bd. III, 241-251

Matzner, E. und B. Ulrich 1981: Bilanzierung jährlicher Elementflüsse in Waldökosystemen im Solling. Z.Pflanzenernähr. Bodenkde. 144, 660-681

Matzner, E., P.K. Khanna, K.J. Meiwes, M. Lindheim, J. Prenzel und B. Ulrich 1982: Elementflüsse in Waldökosystemen im Solling - Datendokumentation - Gött.Bodenkdl.Ber. 71, 1-267

Matzner, E. 1983: Balances of element fluxes within different ecosystems impacted by acid rain. In: B. Ulrich and J. Pankrath: Effects of accumulation of air pollutants in forest ecosystems. D.Reidel Publishing Company, 147-155

Matzner, E. and B. Ulrich 1985: Implications of the chemical soil conditions for forest decline. Experientia 41, 578-584

Matzner, E. and B. Ulrich 1983: The turnover of protons by mineralization and ion uptake in a beech (Fagus silv.) and a Norway spruce ecosystem. In: B. Ulrich and J. Pankrath: Effects of accumulation of air pollutants in forest ecosystems. D. Reidel Publishing Company, 1983, 147-156

Matzner, E. 1984: Annual rates of deposition of polycyclic aromatic hydrocarbons in different forest ecosystems. Water, Air and Soil Pollution 21, 425-434

Mayer, R. 1983: Interactions of forest canopies with atmospheric constituents: Aluminum and heavy metals. In: B. Ulrich and J. Pankrath (eds.): Effects of accumulation of air pollutants in forest ecosystems. D. Reidel Publishing Company, 1983, 47-55

Meyer, F.H. 1984: Mykologische Beobachtungen zum Baumsterben. AFZ, 212-228

Miller, H.G. and Miller, H.D. 1980: Collection and retention of atmospheric pollutants by vegetation. In: B. Ulrich and J. Pankrath: Effects of accumulation of air pollutants in forest ecosystems. D.Reidel Publishing Company, 33-40

Rost-Siebert, K. 1983: Aluminium-Toxizität und -Toleranz an Keimpflanzen von Fichte (Picea abies Karst.) und Buche (Fagus sylvatica L.). Allg.Forstz. 686-689

Schaller, G. und W.R. Fischer 1985: PH-Änderungen in der Rhizosphäre von Mais- und Erdnußpflanzen. Z.Pflanzenernähr. Bodenkde. 184, 306-320

Schimpff, E. 1983: Waldsterben infolge hoher Schadstoffkonzentrationen im Nebel ? Staub-Reinh. Luft 43, 240

Seibt, G. 1983: Zum Einfluß von Luftverunreinigungen auf das Waldwachstum. Vortrag Tagung Sektion Ertragskunde in Neuhaus/Solling 1983

Staaf, H. 1982: Plant nutrient changes in beech leaves during senescence as influenced by site characteristics, Acta Ecol. Plant 3, 161-170

Ulrich, B., U. Steinhardt und A. Müller-Suhr 1973: Untersuchungen über den Bioelementgehalt in der Kronentraufe. Göttinger Bodenkdl.Ber. 29, 133-192

Ulrich, B., R. Mayer und P.K. Khanna 1979: Die Deposition von Luftverunreinigungen und ihre Auswirkungen in Waldökosystemen im Solling. Schriften aus der Forstl.Fak.d.Univ.Göttingen, Bd. 58, Sauerländer-Verlag

Ulrich, B. und E. Matzner 1983: Abiotische Folgewirkungen der weiträumigen Ausbreitung von Luftverunreinigungen. Forschungsbericht 104 02 615, UBA Berlin

Ulrich, B. und E. Matzner 1983: Ökosystemare Wirkungsketten beim Wald- und Baumsterben. Forst- und Holzwirt 18, 468-474

Ulrich, B. 1983: Interaction of forest canopies with atmospheric constituents: SO_2, alkali and earth alkali cations and chloride. In: B. Ulrich and J. Pankrath: Effects of accumulation of air pollutants in forest ecosystems. D. Reidel Publishing Company, 1983, 33-45

Ulrich, B. 1983a: A concept of forest ecosystems stability and of acid deposition as driving force for destabilization. In: B. Ulrich and J. Pankrath: Effects of accumulation of air pollutants in forest ecosystems. D.Reidel Publishing Company, 1983, 1-29

Ulrich, B., R. Mayer and P.K. Khanna 1980: Chemical changes due to acid precipitation in a loess derived soil in Central Europe, Soil Science 130, 193-199

Wisniewski, J. 1982: The potential acidity associated with dews, frosts and fogs. Water, Air and Soil Pollution 17, 361-377

Zech, W. und E. Popp 1983: Magnesiummangel, einer der Gründe für das Fichten- und Tannensterben in NO-Bayern. Forstwiss.Ctbl. 102, 50-55

Zöttl, H.W. und E. Mies 1983: Die Fichtenerkrankung in den Hochlagen des Südschwarzwaldes. Allg. Forst- und J.Ztg. 154, 110-114

PROCESSES IN THE CANOPY OF TREES : INTERNAL AND EXTERNAL TURNOVER OF ELEMENTS

Godt,J., M. Schmidt and R. Mayer
Gesamthochschule Kassel, FB Stadt- und Landschafts-
planung, Abt. Landschaftsökologie/Bodenkunde
Henschelstr. 2
D-35oo Kassel
West-Germany

Abstract. To estimate intake of air pollutants and nutrients by balancing the elemental turnover in the canopy of tree stands requires knowledge of internal (leaching) and external (atmospheric intake) fluxes. Wash-off experiments with twigs from spruce differently exposed to atmospheric pollution show a large variety in the contribution of interception deposition (aerosols and gases) to total atmospheric input. By using a gas chamber with filtered air, additional information on plant uptake and leaching (beech) could be gained. It can be concluded from these experiments that plant uptake (and thereby leaching) of Pb and Cr can be neglected, whereas plant uptake of Cd and Cu will be very small. As to Mn, Ca and Mg, plant uptake and thereby leaching plays a more important role. This has to be taken into account, when balancing the turnover of metals in the canopy of tree stands.
High concentrations of potentially toxic heavy metals on vegetation surfaces, especially after longer periods of dry deposition, may lead to leaf damages. These processes may be an additional factor contributing, among others, to forest decline.

1. Introduction

Since forest decline has become more and more obvious, determination of the atmopsheric input of toxic substances and nutrients has gained special interest. Potentially toxic substances such as S and heavy metals are thought to contribute, to a larger extend, to forest decline (Rat der Sachverständigen für Umweltfragen 1984). As the amount of intake of aerosols and gases is strongly correlated with the physical properties of the canopy surface, Ulrich et al. (1979) developed a model in order to balance the turnover of elements in the canopy of forest stands. From these balances predictions of internal fluxes (plant uptake and leaching) and external fluxes (atmospheric input) can be derived. In these balances the element flux connected with cannopy drip can be defined as follows :

(1) BN = ND + ID

BN = canopy drip (Bestandesniederschlag)
ND = deposition in the open field (Freiflächendeposition)
ID = deposition by interception of gases and aerosols (Interceptionsdeposition)
Q = sink / source term (Quellen-/Senkenterm)

Retention of atmospheric substances in the canopy leads to a negative value of Q (sink), leaching of substances of internal turnover leads to a positive value (source). For some elements the source/sink term can be neglected (Q = 0 for Na, Cl, S). Thus for these elements ID, which can not be measured directly on vegetation surfaces, can be estimated with good approximation from eq. (1) by determining BN - ND = ID (Ulrich et al. 1979, Meiwes et al. 1984, Mayer 1985).
As to the heavy metals, the problem arises that not only leaching of the canopy may play a role but, in addition to this, the canopy can act as a sink by ad/absorption of elements after deposition. The aim of our experiments was to distinguish internal and external turnover of trace elements in order to establish balances from which total deposition and interception deposition can be estimated.

2. Methods

The contribution of dry deposition to the element load of the canopy drip was estimated from wash-off experiments with twigs from spruce (Picea abies L.) trees taken out of forest stands in Southern Lower Saxony with various levels of atmospheric pollution. Sampling of the twigs (seventh whirl of dominating trees, about 2oo g dry material) was done at different trees or parts of trees with respect to wind exposure, position in the stand and in the landscape. From these variations different inputs by atmospheric deposition could be expected and thereby different elemental behaviour could be studied. In the laboratory the twigs were sprayed with deionized water brought to pH 4.o by addition of HNO_3. Differences in the amounts of elements washed off were attributed to dry atmospheric input during the previous dry period. The wash-off solution was collected in succecive fractions and analysed to simulate the processes of wash-off under natural conditions.

Another experiment was carried out in the experimental site of the Solling area (beech, Fagus sylvatica L.). Single twigs, still being connected to the tree, have been carefully wrapped in a plastic cover and been gased with filtered air (aerosolfilter retaining particles > o.o1 μm, fig.1). These and unprotected reference twigs have been sprayed in the way described above after dry periods of various duration. In addition, leaves from protected and unprotected twigs have been taken during the vegetational period and analyzed for elemental content.

Fig. 1 : Experimental set up

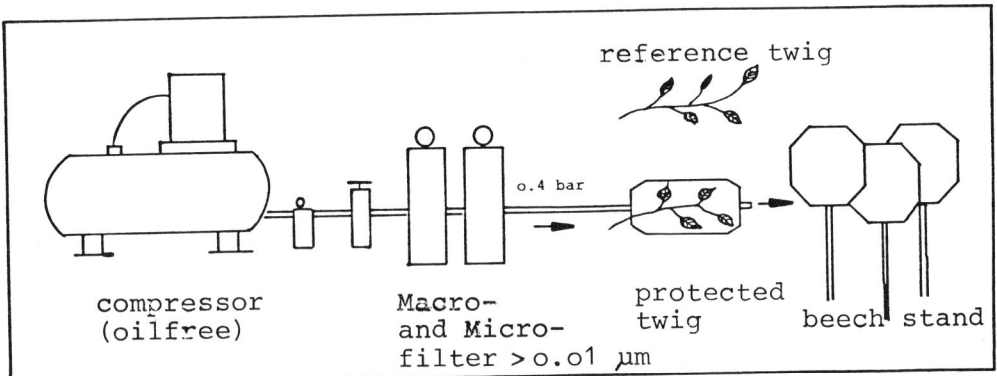

3. Results

In fig. 2 the results of the wash-off experiments with twigs from a spruce stand (about 2o years old) in the neighbourhood of a smelting factory (Oker-Harlingerode) are shown. The twigs have been taken into laboratory after a 7-days period of dry weather. It is obviousthat especially in the second wash-off fraction (1o-5o ml) extremely high concentrations of heavy metals have been measured on those twigs, which have been taken from the edge of the stand (Cd : 35 μg/l, Pb 92o μg/l, Cu 2oo μg/l, Zn 1.63 mg/l, Mn o.23 mg/l). At the same time extremely low pH-values have been measured (minimum 2.3 pH). Comparison of the quantities washed-off (calculated on a comparable basis of 2oo g dry material, table 1) shows higher quantities of heavy metals on the surface of those twigs, which have been taken from the exposed edge of the stand, while Ca, Mn and Mg don´t show such distinct differences. After the wash-off experiment with deionized water the twigs have been sprayed with diluted acid (o.65 % HNO_3, 5o ml) in order to mobilize acid-soluble metals. In this experiment considerable quantities

Fig. 2 : Wash-off experiment Oker-Harlingerode
(30.5.1982), spruce
+ —— + = edge of the stand
• —— • = inside the stand

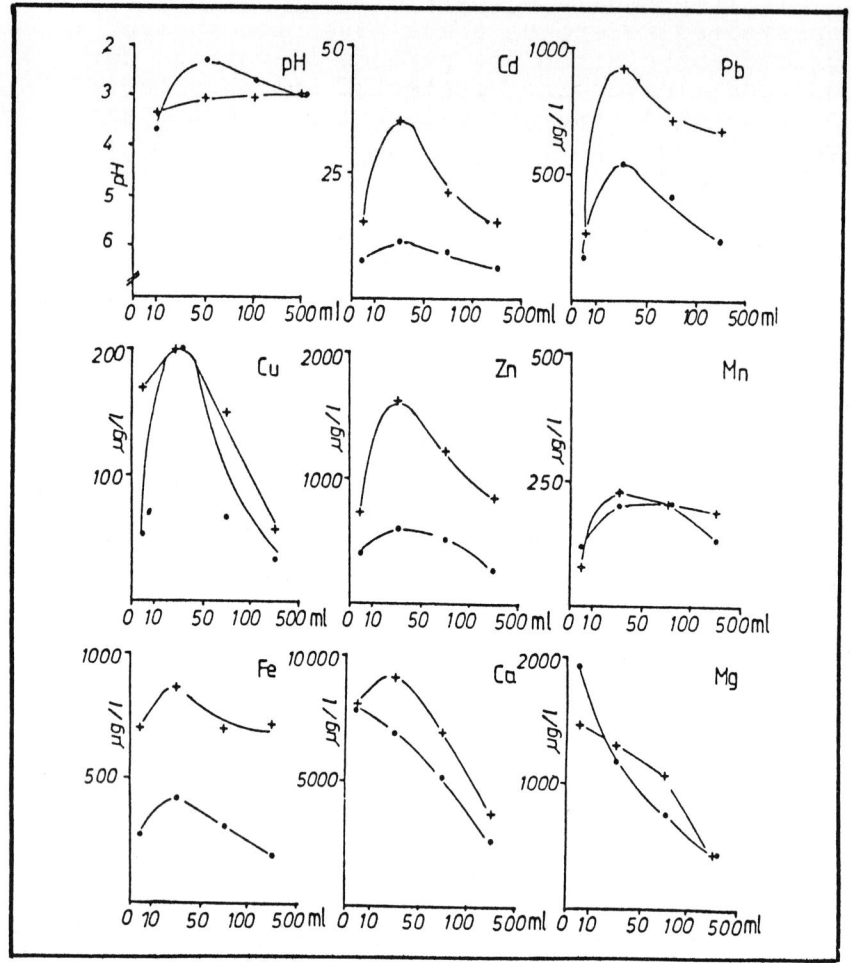

Tab. 1 : Comparison of wash-off quantities,
Oker-Harlingerode (30.5.1982), spruce

		Cd	Pb	Cu	Zn	Mn	Fe	Ca	Mg
		µg/200 g dry m.							
(1)	0-500 ml inside of the stand	3.5	142	27	170	90	110	1770	290
(2)	0-500 ml edge of the stand	8.9	355	42	490	100	380	2350	300
	(1) in % of (2)	40	40	63	35	88	29	75	96

Fig. 3 : Wash-off experiment Solling F 1 beech
+ ———— + 4.4.1982, after dry period
. ———— . 12.4.1982, after several showers

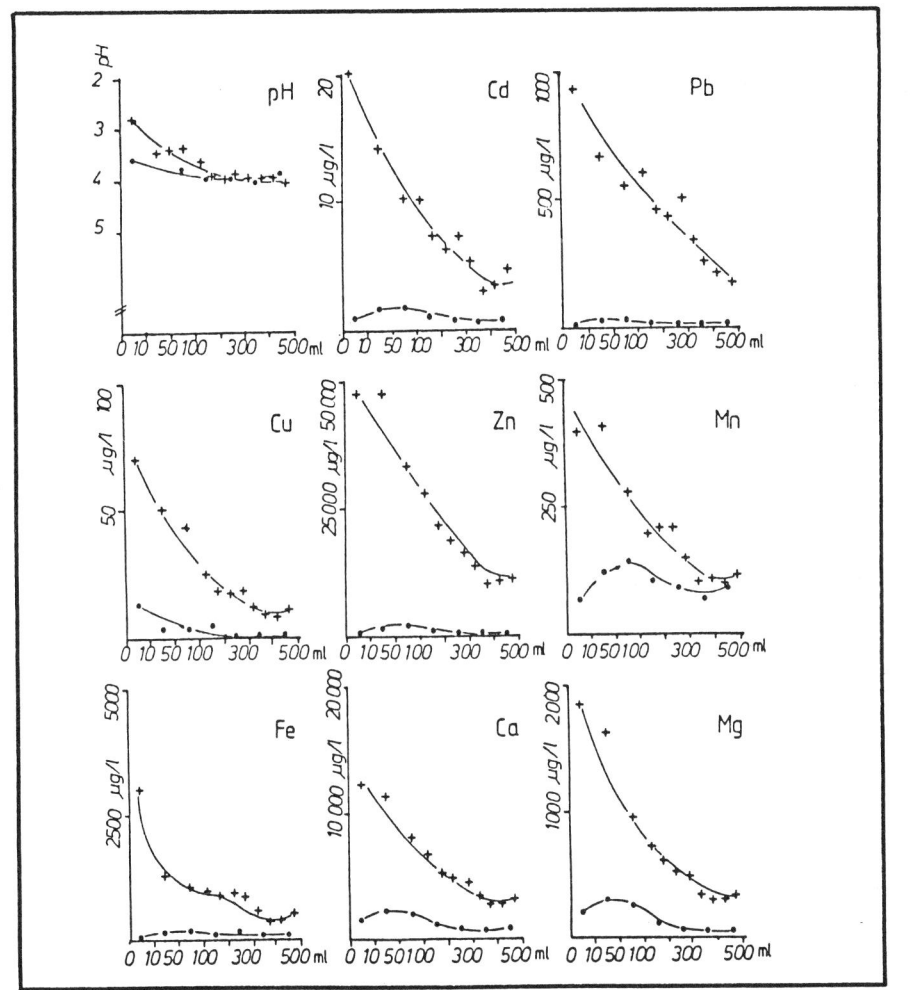

Tab. 2 : Comparison of wash-off quantities after a dry period and several showers (0-500 ml)

	Cd	Pb	Cu	Zn	Mn	Fe	Ca	Mg
				µg/200 g dry m.				
after several showers	1.0	13	2.1	2050	110	230	560	110
after dry period	7.7	270	11.2	1340	120	550	331	400

of Pb and Fe have been washed off.
In fig. 3 the results of a wash-off experiment with twigs of spruce from the Solling area are shown. The twigs have been taken after a long period (about twenty days) of dry weather and, in contrast to this, after several showers (7.4.-12.4.1982 : 21 mm). The results show clearly that the twig exposed during a long period of dry deposition (table 2) yielded much higher amounts of most metals -except Mn - in the wash-off solution than that sprayed after several showers. In table 3 the concentrations in the fraction o-1o ml are compared with those measured in the canopy drip of the spruce stand (for several years) (Mayer 1981, Matzner et al. 1984). By this comparison it is obvious, that the concentrations of heavy metals and other components are much lower than those which can be expected on the vegetation surface.

Tab. 3 : Comparison of concentrations in wash-off liquids (0-10 ml) and canopy drip (Mayer 1981, Matzner et al. 1984), spruce, Solling F 1

	pH	Cd	Pb	Cu	Mn	Fe	Ca	Mg
		---------- µg / 1 ------------
wash-off experiment 0-10 ml 4.4.1982	2.9	20.2	935	71	400	3030	1240	2340
canopy drip	3.5	3.4	78	34	880	300	4000	800

In table 4 the results of wash-off experiments with twigs from different stands (BAB A 7: highway, Solling, F 1: rural area, Hardegsen: on the leeside of a lime work) are shown (approximately same time of harvest). Different site

Tab. 4 : Comparison of wash-off quantities from spruce twigs (0-500 ml/200 g dry m.) from different locations

	Cd	Pb	Cu	Zn	Mn	Fe	Ca	Mg
	---------- µg / 200 g dry m. --------
BAB A 7 highway Hannover-Kassel 5.4.1982	1.8	337	37	310	260	640	1030	1290
Solling, F 1 4.4.1982	7.7	270	11	1340	120	550	3300	400
Hardegsen 6.4.1982	2.2	163	30	240	260	620	19200	2510

Fig. 4 : Concentrations of metals in wash-off liquids of twigs, protected and unprotected from atmospheric deposition, beech (Fagus sylvatica L.), Solling B 1

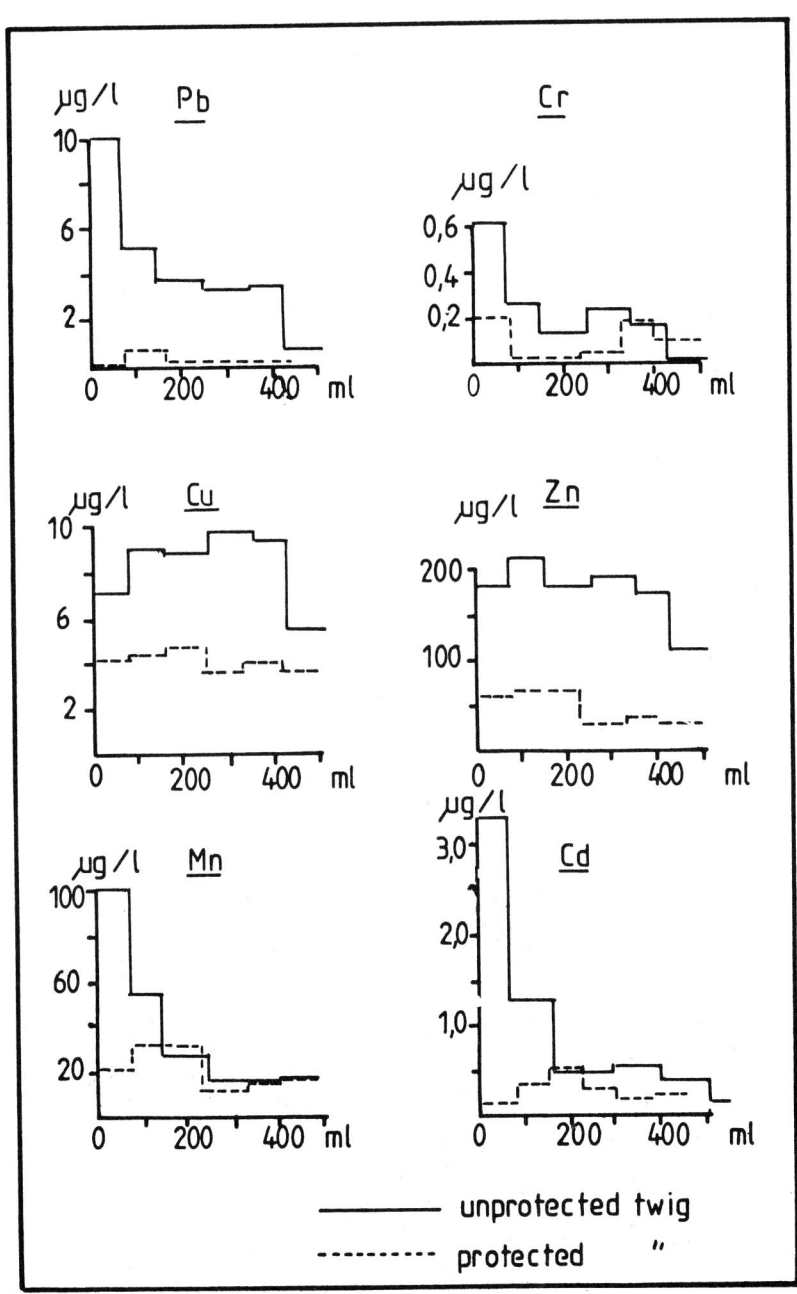

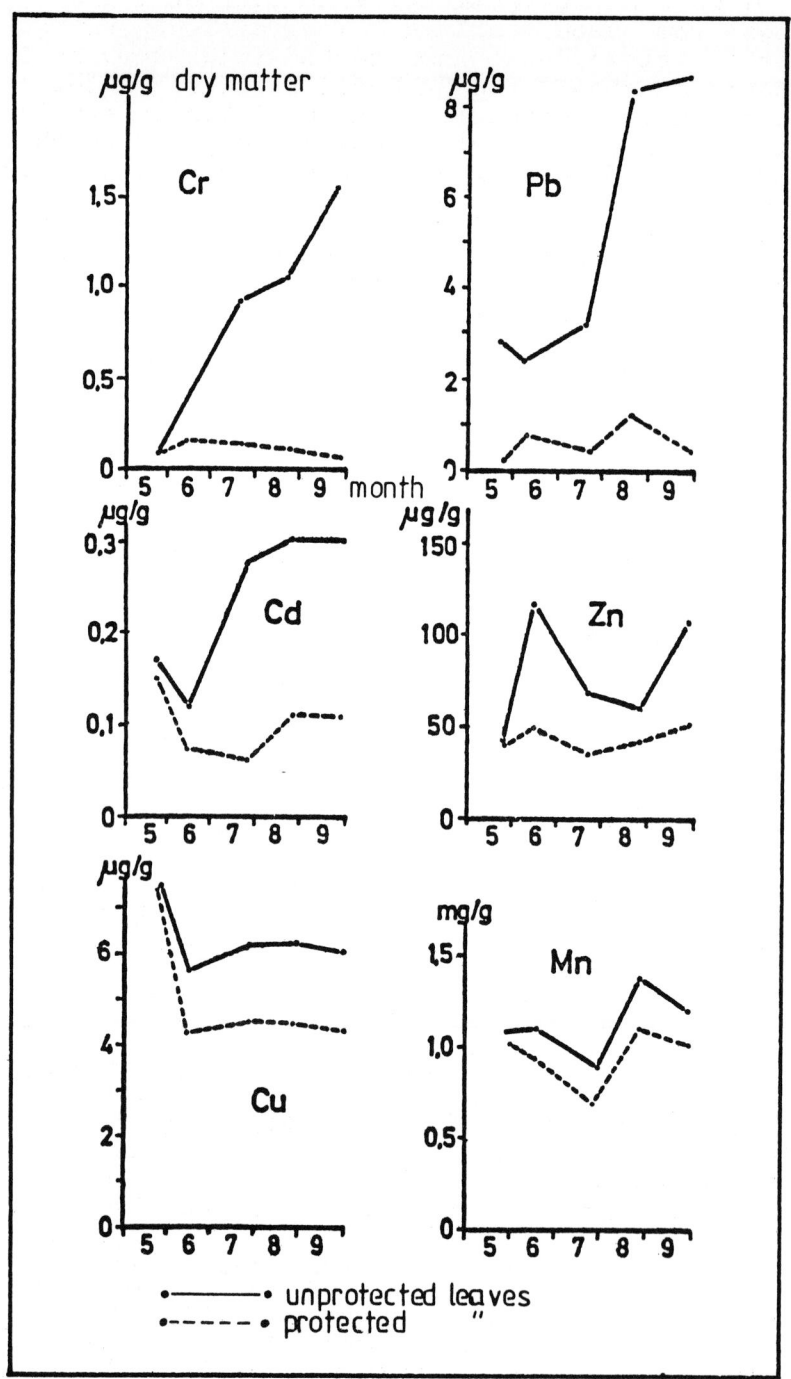

Fig. 5 : Concentration of metals in beech (Fagus sylvatica L.) leaves, protected and unprotected from atmospheric deposition, Solling B 1, May to Oct. 1984

conditions are reflected : BAB A 7 : high amounts of Pb from vehicle exhausts, Solling : high Cd and Zn values, probably from long range atmospheric transport, Hardegsen : high Ca and Mg quantities from lime work.

Results from wash-off experiments with protected and unprotected twigs from beech (Solling B 1) are given in Fig. 4. It is clearly visible, that unprotected twigs show by far higher concentrations in the wash-off solutions after a longer period of dry deposition. The differences are larger for the elements Pb, Cr and Cd, while, compared to these elements, Cu, Zn and Mn show high concentrations also in the unprotected twig.

Fig. 5 shows the heavy metal content of beech leaves during a vegetation period. While Cr, Pb, and Cd show accumulation in the protected leaves the Cu, Zn and Mn contents do not increase over the vegetation period. In protected leaves only very low concentrations of Pb and Cr have been found, concentration remains more or less constant. Some elements (Cd, Cu and Mn) even show a decrease in the beginning of the vegetation period. This seems to be a "dilution effect" during spring, when organic matter is built up faster than metal elements are supplied by root uptake or atmospheric deposition.

4. Discussion

4.1. Interpretation of the results

In the wash-off experiments the role of interception depositiion could be pointed out for heavy metals and macroelements. Differences could be shown up which seem to be specific for single elements. The wash-off experiments showed clearly that the heavy metals considered, Cd, Pb and - with some limitation - Cu could be washed off in higher quantities from those plant parts which were exposed to interception deposition. By further experiments it could be shown that Pb is taken up by plants in quasinegligible amounts while Cd and Cu are taken up in a very small amount (Mayer 1981, Godt 1986, Schmidt and Schultz 1985). Thereby only very small leaching of Cd and Cu from the canopy can be expected. This means that the wash-off qunatities of Pb, Cd and Cu are reflecting the variation in interception deposition depending upon surface structure and physical environment (f.e. micrometeorology of the trees).

In contrast to the heavy metals Pb, Cd and Cu the macroelements Mn, Ca and Mg show a different behaviour. These elements which are known to be more mobile in the biogeochemical cycle (Ernst und Josse van Damme 1983), don´t show distinctly higher wash-off quantities from exposed plant parts. This means that these elements are, in the canopy drip, deriving to a larger extent from the internal cycle (leaching).

Experiments concerning the physical properties of aerosols
(Höfken and Gravenhorst 1980, Schmidt et al. 1985, Godt et
al. 1985, Godt 1986) have shown that, following the greater
mass median diameter of aerosols, sedimentation will play
a more important role in the case of Mn, Ca and Mg than for
Cd, Pb, Cu and Zn (here interception deposition is more
important).
The statement, that the heavy metals Pb and Cr are deriving
to only very small amounts from the internal cycle, is fully
supported by observations from experiments with protected
and unprotected twigs from beech. These elements (including
Cr) showed smaller wash-off qunatities and smaller total
concentrations in the "protected" leaves. The smallest concentration differences between protected and unprotected
leaves shows Mn which gives evidence that this element de-
rives mainly from internal turnover.
Neglecting plant uptake (via roots), total deposition of Pb
and Cr can be estimated by adding the deposition rate in
canopy drip (liquid) and litterfall (solid) and the rate
of increase in biomass (wood) storage.
The elements Cu and Zn are deriving from the internal cycle
to a smaller degree; thereby total input of these elements
can be estimated according to the method described by
Mayer (1981).
For all elelements to which internal turnover contributes
to a large extent to the flux in the canopy drip (Mn, Mg,
Ca and other major elements) total deposition can be estimated by using "guiding elements" like Na, S or Cl (see
above), as described by Ulrich et al. (1979), Meiwes et al.
(1984) and Mayer (1985).

4.2. Toxicological aspects

The high concentration of heavy metals in the initial wash-
off fractions are reflecting high concentrations on vegetational surfaces. In this context, the very low pH in combination with high concentrations of heavy metals (high
solubility and thereby mobility) may point at potentially
toxic effect. It can be expected that the concentrations
of heavy metals on the vegetation surface will be considerably higher than those measured in the wash-off experiments.
Higher concentrations may occur after longer dry periods in
a following dissolution phase (for example morning dew).
In this phase the dissolved toxic substances may rinse the
surface and accumulate on the tip of a leaf when drying off
again (see also Klemm und Frevert 1985).
Damage to adjecent plant cells may be the consequence, if
the tree is unable to buffer the water film and to immobi-
lyze the heavy metals.

References

Chamberlain, A. C. (1955): Aspects of travel and deposition of aerosols and vapour clouds. Atomic Energy Research Establishment, AERE HP/R 1261, Harwell

Ernst, W.H.O. und Josse van Damme (1983): Umweltbelastung durch Mineralstoffe. Gustav Fischer Verlag Stuttgart

Godt, J., R. Mayer und B. Georgi (1985): Die Interception von Schwermetallen in Abhängigkeit von der Topographie und der Orographie . VDI-Kolloquium "Waldschäden - Einflußfaktoren und ihre Bewertung", 18.-2o. Juni 1985 Goslar, VDI-Bericht 56o, 333-356

Godt, J. (1986): Untersuchung von Prozessen im Kronenraum von Waldökosystemen und deren Berücksichtigung bei der Erfassung von Schadstoffeinträgen - unter besonderer Beachtung der Schwermetalle , Diss. an der Gesamthochschule Kassel , Berichte des Forschungszentrums Waldökosysteme/Waldsterben 19, Göttingen, 264 S.

Höfken, K.D. and G. Gravenhorst (198o): Concentration and size distribution of trace substances in aerosols above and beneath a forest canopy. Tagung Gesellsch. f. Aerosolforschung Schmallenberg, Okt. 198o

Höfken, K.D. (1983): Input of acidifiers and heavy metals to a German forest area due to dry and wet deposition. In : B. Ulrich and J. Pankrath (eds.): Effects of Accumulation of Air Pollutants in Forest Ecosystems, D.Reidel Publishing Company, 57-64

Klemm, O. und T. Frevert (1985): Säure- und redoxchemisches Verhalten von Niederschlagswasser beim Abdampfen von Oberflächen. VDI-Kolloquium "Waldschäden - Einflußfaktoren und ihre Bewertung," 18.-2o. Juni 1985 Goslar, VDI-Bericht 56o, 457-465

Matzner, E., P. K. Khanna, K.J. Meiwes, E. Cassens-Sasse, M. Bredemeier und B. Ulrich (1984): Ergebnisse der Flüssemessungen in Waldökosystemen. Berichte des Forschungszentrums Waldökosysteme/Waldsterben Göttingen 2, 29-49

Mayer,R. (1985): Verfahren zur Erfassung der Schadstoffzufuhr in Waldökosystemen. Staub Reinhaltung der Luft 45, Nr. 6, Juni 1985, 267-292

Mayer,R. (1981): Natürliche und anthropogene Komponenten des Schwermetallhaushalts von Waldökosystemen . Göttinger Bodenkundliche Berichte 7o, 1 - 292

Meiwes, K.J., M. Hauhs, H. Gerke, N. Asch, E. Matzner und N. Lamersdorf (1984): Chemische Untersuchungsverfahren für Mineralboden, Auflagehumus und Wurzeln zur Charakterisierung und Bewertung der Versauerung in Waldböden . Berichte des Forschungszentrums Waldökosysteme/Waldsterben 7, Göttingen, 142 S.

Rat der Sachverständigen für Umweltfragen (1983) : Waldschäden und Luftverunreinigungen, Sondergutachten März 1983, Verlag W. Kohlhammer Stuttgart und Mainz

Ulrich, B., R. Mayer und P.K. Khanna (1979): Deposition von Luftverunreinigungen und ihre Auswirkungen in Waldökosystemen des Solling. Schriften der Forstlichen Fakultät der Universität Göttingen und Nds. Forstl. Versuchsanstalt 58, 291 S.

Ulrich, B. (1983): Interaction of forest canopies with atmospheric constituents : SO_2, alkali and earth alkali cations and chloride. In: B. Ulrich and J. Pankrath (eds.): Effects of accumulation of air pollutants in forest ecosystems. D. Reidel Publishing Company Dordrecht

Schmidt, M., R. Mayer und B. Georgi (1985): Beeinflussung der Interceptionsdeposition (Trockene Deposition) durch die Aerosolkonzentration und -größenverteilung in einem Buchenbestand (Solling), VDI-Kolloquium "Waldschäden - Einflußfaktoren und ihre Bewertung" 18.-2o. Juni 1985 Goslar, VDI-Bericht 56o 423-438

Schmidt, M. und R. Schultz (1985): Dry deposition of heavy metals in a beech stand. Proc. Int. Conf. Heavy Metals in the Environment Athens, 5o6-5o8

LIST OF PARTICIPANTS

Amtmann, R.
Institut für Bioklimatologie
und angewandte Meteorologie
Amalienstr. 52/III
8000 München 40

Asman, W.
Rijksuniversiteit Utrecht
Instituut meteorologie en
oceanografie
Princetonplein 5
2506 Utrecht
Niederlande

Bächmann, Prof. Dr. K.
Technische Hochschule Darmstadt
FB Anorganische Chemie und Kernchemie
Hochschulstr. 4
6100 Darmstadt

Bär, F.
Kernforschungszentrum Karlruhe
Institut für Meteorologie
und Klimaforschung
7500 Karlsruhe

Balasz, Dr. A.
Institut für Forsthydrologie der
Hess. Forstl. Versuchsanstalt
Prof. ölkersstr. 6
3510 Hann. Münden

Ballach, Dr. H.J.
Gesamtverband des deutschen
Steinkohlebergbaus
Friedrichstr. 1
4300 Essen

Baumbach, Dr. G.
Institut für Verfahrenstechnik
und Dampfkesselwesen
Universität Stuttgart
Pfaffenwaldring 23
7000 Stuttgart

Beheng, Dr. K.
Institut für Meteorologie und
Geophysik
Universität Frankfurt
Feldbergstr. 47
6000 Frankfurt am Main 1

Beilke, Dr. S.
Umweltbundesamt
Pilotstation Frankfurt
Frankfurter Str. 135
6050 Offenbach

LIST OF PARTICIPANTS

Beltz, N.
Referat für Umweltschutz
Universität Frankfurt
Robert-Mayerstr.
6000 Frankfurt/Main

Bingemer, Dr. H.
Max-Planck-Institut
für Chemie
Abt. Luftchemie
Saarstraße 23
6500 Mainz

Blank, Dr. L. W.
GSF - München
Expositionsanlage
8042 Neuherberg

Bredemaier, M.
Institut f. Bioklimatologie
Büsgenweg 1
3400 Göttingen

Brechtel, Prof. Dr. H.M.
Institut für Forsthydrologie der
Hess. Forstl. Versuchsanstalt
Prof. ölkersstr. 6
3510 Hann. Münden

Bockholt, Dr. B.
Landesamt für Umweltschutz
und Gewerbeaufsicht
Rheinland - Pfalz
Postfach 119
6504 Oppenheim

Borchert, Dr. H.
Landesamt für Umweltschutz
und Gewerbeaufsicht
Rheinland - Pfalz
Postfach 119
6504 Oppenheim

Bucher, P.
Universität Zürich
Anorganisches Institut
Winterthurer Str. 190
8057 Zürich
Schweiz

Christ, W.
Referat für Umweltschutz
Universität Frankfurt
Robert-Mayerstr.
6000 Frankfurt/Main

Deumling, Dr. D.
Schloss Schönstein
5248 Wissen/Sieg

LIST OF PARTICIPANTS

Doms, G.   Institut für Meteorologie und
                            Geophysik
                            Universität Frankfurt
                            Feldbergstr. 47
                            6000 Frankfurt am Main 1

Dröscher, F.   Institut für Verfahrenstechnik
                            und Dampfkesselwesen
                            Universität Stuttgart
                            Pfaffenwaldring 23
                            7000 Stuttgart 80

Enders, Dr. G.   Institut für Bioklimatologie
                            und angewandte Meteorologie
                            Universität München
                            Amalienstr. 52/III
                            8000 München 40

Fangmeier, A.   Institut für Pflanzenökologie
                            Heinrich-Buff-Ring 38
                            6300 Gießen

Fiedler, Prof. Dr. F.   Institut für Meteorologie und
                            Klimaforschung
                            Kernforschungszentrum Karlsruhe
                            Kaiserstr. 12
                            7500 Karlsruhe

Fuchs, G.   Technische Hochschule Darmstadt
                            FB Anorganische Chemie und Kernchemie
                            Hochschulstr. 4
                            6100 Darmstadt

Fuhrer, W.   Eidgenössische Anstalt für Wasser-, Abwasser-
                            und Gewässerschutz EAWAG
                            Überlandstr. 133
                            CH 8600 Dübendorf
                            Schweiz

Gerdsmeier, G.   Biologisches Institut
                            Abt. Zoologie
                            Universität Münster
                            4400 Münster

Georgii, Prof. Dr. H. W.   Institut für Meteorologie und
                            Geophysik
                            Universität Frankfurt
                            Feldbergstr. 47
                            6000 Frankfurt am Main 1

Gies, Prof. Dr. T.	Institut für Biologie Abt. Didaktik Universität Frankfurt Sophienstr. 1 6000 Frankfurt am Main 1
Gietl, G.	Bayer. Forstliche Versuchs- und Forschungsanstalt Schellingstr. 12 - 14 8000 München 40
Glatzel, Prof. Dr. G.	Institut für Forstökologie Universität für Bodenkultur Peter Jordan Str. 82 A-1190 Wien Austria
Godt, Dr. J.	Gesamthochschule Kassel FB Stadt- und Landschaftsplanung Abt. Landschaftsökologie/Bodenkunde Henschelstr. 2 3500 Kassel
Gravenhorst, Prof. Dr. G.	Institut f. Bioklimatologie Büsgenweg 1 3400 Göttingen
Grosch, S.	Institut für Meteorologie und Geophysik Universität Frankfurt Feldbergstr. 47 6000 Frankfurt am Main 1
Hanewald, Dr. K.	Hessische Landesanstalt für Umwelt Unter den Eichen 6200 Wiesbaden
Hahn, Dr. J.	Max-Planck-Institut für Chemie Abt. Luftchemie Saarstraße 23 6500 Mainz
Harssema, H.	Agricultural University Wageningen Department of Air Pollution P. O. Box 8129 6700 EV Wageningen The Netherlands

LIST OF PARTICIPANTS

Herbert, Prof. Dr. F.	Institut für Meteorologie und Geophysik Universität Frankfurt Feldbergstr. 47 6000 Frankfurt am Main 1
Hertz, J.	Universität Zürich Anorganisches Institut Winterthurer Str. 190 8057 Zürich Schweiz
Höfken, Dr. K.-D.	Gesellschsft für Strahlenforschung - BPT - Josephspitalstr. 15 8000 München 2
Hofschreuder, P.	Agricultural University Wageningen Department of Air Pollution P. O. Box 8129 6700 EV Wageningen The Netherlands
Jaeschke, Dr. W.	Referat für Umweltschutz Universität Frankfurt Robert-Mayerstr. 6000 Frankfurt/Main
Johnson, A.	Eidgenössische Anstalt für Wasser-, Abwasser- und Gewässerschutz EAWAG Überlandstr. 133 CH 8600 Dübendorf Schweiz
Kalte, B.	Institut für Meteorologie und Geophysik Universität Frankfurt Feldbergstr. 47 6000 Frankfurt am Main 1
Kazda, Dr. M.	Institut für Forstökologie Universität für Bodenkultur Peter Jordan Str. 82 A-1190 Wien Austria
Kuttler, Dr. W.	Universität Bochum Geographisches Institut Postfach 102148 4630 Bochum

LIST OF PARTICIPANTS

Kraemer,   Kernforschungsanlage Jülich
                            ICH 3
                            Postfach 1913
                            5170 Jülich

Kramm, G.   Industrieanlagen- und Betriebs-
                            gesellschaft mbH
                            Einsteinstr. 20
                            8012 Ottobrunn

Lehnardt, Dr. F.   Institut für Forsthydrologie
                            der Hess. Forstl. Versuchsanstalt
                            Prof. Oelkerstr. 6
                            3510 Hann. Münden 1

Lüpkes, C.   Institut für Meteorologie und
                            Geophysik
                            Universität Frankfurt
                            Feldbergstr. 47
                            6000 Frankfurt am Main 1

Matzner, Dr. E.   Forschungszentrum Waldökosysteme
                            Universität Göttingen
                            Büsgenweg 2
                            3400 Göttingen

Mayer, Prof. Dr. R.   Gesamthochschule Kassel
                            FB Stadt- und Landschaftsplanung
                            Abt. Landschaftsökologie/Bodenkunde
                            Henschelstr. 2
                            3500 Kassel

Müller, Dr. J.   Umweltbundesamt
                            Pilotstation Frankfurt
                            Frankfurter Str. 135
                            6050 Offenbach

Nestlen, Dr. N.   Institut f. Bioklimatologie
                            Büsgenweg 1
                            3400 Göttingen

Ober, E.   Technische Universität Wien
                            Getreidemarkt 9
                            A-1060 Wien
                            Austria

Perseke, Dr. C.   Deutscher Wetterdienst
                            Abt. K3
                            Frankfurter Str. 135
                            6050 Offenbach

LIST OF PARTICIPANTS

Puxbaum, Dr. H.	Technische Universität Wien Getreidemarkt 9 A-1060 Wien Austria
Queirolo, N.	Institut für Chemie ICH 4 Kernforschungsanlage Jülich GmbH Postfach 913 5170 Jülich
Rall, A.	Institut für Bioklimatologie und angewandte Meteorologie Aventinstr. 14/III 8000 München 5
Rohbock, Dr. E.	Battelle Institut e.V. Am Römerhof 35 6000 Frankfurt am Main
Schaub, Prof. Dr. H.	Botanisches Institut Universität Frankfurt Siesmayerstr. 70 6000 Frankfurt am Main
Schmidt, M.	Gesamthochschule Kassel FB Stadt- und Landschaftsplanung Abt. Landschaftsökologie/Bodenkunde Henschelstr. 2 3500 Kassel
Schmitt, G.	Institut für Meteorologie und Geophysik Universität Frankfurt Feldbergstr. 47 6000 Frankfurt am Main 1
Schönwiese, Prof. Dr. C.	Institut für Meteorologie und Geophysik Universität Frankfurt Feldbergstr. 47 6000 Frankfurt am Main 1
Sigg, L.	Eidgenössische Anstalt für Wasser-, Abwasser- und Gewässerschutz EAWAG überlandstr. 133 CH 8600 Dübendorf Schweiz
Teichmann, U.	Institut für Bioklimatologie und angewandte Meteorologie Universität München Amalienstr. 52/III 8000 München 40

Thomas, J.   Referat für Umweltschutz
                            Universität Frankfurt
                            Robert-Mayerstr.
                            6000 Frankfurt/Main

Trefz-Malcher, G.   Forstl. Versuchs- und Forschungs-
                            anstalt
                            Baden-Württemberg
                            Abt. Botanik- und Standortskunde
                            Fasanengarten
                            7000 Stuttgart 31

Valenta, Dr. P.   Institut für Chemie ICH 4
                            Kernforschungsanlage Jülich GmbH
                            Postfach 913
                            5170 Jülich

Vermetten, A. W. M.   Agricultural University Wageningen
                            Department of Air Pollution
                            P. O. Box 8129
                            6700 EV Wageningen
                            The Netherlands

Wallenwein, H.   Institut für Meteorologie und
                            Geophysik
                            Universität Frankfurt
                            Feldbergstr. 47
                            6000 Frankfurt am Main 1

Warneck, Prof. Dr. P.   Max-Planck-Institut
                            für Chemie
                            Abt. Luftchemie
                            Saarstr. 23
                            6500 Mainz

Wegener, Dr. H.-R.   Institut für Bodenkunde und
                            Bodenhaltung
                            der Justus-Liebig Universität
                            Gießen
                            Wiesenstr. 3-5
                            6300 Gießen

Winkler, Dr. P.   Deutscher Wetterdienst
                            Meteorologisches Observatorium Hamburg
                            Frahmredder 95
                            2000 Hamburg 65

Wittig, R.   Universität Düsseldorf
                            Abteilung Geobotanik
                            4000 Düsseldorf

LIST OF PARTICIPANTS

Wobrock, W.
: Institut für Meteorologie und
Geophysik
Universität Frankfurt
Feldbergstr. 47
6000 Frankfurt am Main 1

Wolkewitz, N.
: Institut für Bodenkunde und
Bodenerhaltung
der Justus-Liebig Universität
Gießen
Wiesenstr. 3-5
6300 Gießen

Wendt, S.
: Inst. für Umweltschutz
Universität Dortmund
Postfach 500 500
4600 Dortmund 50

Zimmermann, R.
: Institut für Meteorologie und
Geophysik
Universität Frankfurt
Feldbergstr. 47
6000 Frankfurt am Main 1

SUBJECT INDEX

acid deposition	25, 35, 47, 79, 89, 103, 238, 247
acid gases	3, 13, 25, 177, 189
bioindication	223, 235
below cloud scavenging	165
canopy resistance	3
cadmium	
- emission sources	69
- concentration and deposition	35, 47, 69, 79, 129, 155, 201, 215, 235, 263
- in the food web	223
chloride	35, 47, 79, 89, 155, 189, 247
concentration profiles	6, 17, 181
conductivity	129, 143, 165
correlation analyses	25
deposition	
- bulk	25, 35, 47, 79, 89, 247
- dry	35, 89, 102, 215, 201, 238
- gaseous	3, 13, 177
- wet	35, 69, 89, 165, 201, 238
deposition/canopy-interactions	247, 263
deposition velocities	3, 13
elemental turnover	263
episodes of deposition	39
filtering effect	40, 86, 177, 248
fog	
- water collector	111, 129, 143
- physical and chemical properties	112
- temporal distribution	129
- chemical composition	129, 143

forest micrometeorology	13
gaseous concentrations	3, 13, 25, 89, 177, 189, 235
H^+-deposition	82, 172, 249
HCL	189
heavy metals	
- concentration and deposition in rain	52, 69, 155
- water solubiltiy	155, 201
- bioavailability	201
immission rates	25, 89
in cloud scavenging	155, 165.
interception	41, 79, 129, 177, 215, 247, 263,
lead	
- emission sources	69
- concentration and deposition	35, 47, 69, 79, 129, 155, 201, 215, 235, 263
- in the food web	223
litter fall	3, 255, 263,
nitrate	25, 35, 47, 79, 89, 101, 118, 129, 149, 155, 191, 247
NO_x	3, 13, 177
ozone	3, 13, 177
pH	
- in rainwater	44, 81, 92, 133, 155, 165
- in fog	111, 129, 143
phosphorus	85
rainout-fraction	155, 165
rime	101
snow	
- deposition on -surfaces	101
- lysimeter	107
SO_2	3, 13, 25, 177
sequential precipitation sampler	155, 165
stem flow	215

SUBJECT INDEX

sulfate	35, 47, 79, 89, 101, 127, 129, 143, 155, 191, 235
water solubility of heavy metals	162, 201
wet only sampler	35, 69
winter deposition rates	101
throughfall	35, 47, 79, 215, 247, 263